Journal of
Neural
Transmission

Supplementum 37

H. Ågren, J.-L. Martinot, and F.-A. Wiesel (eds.)

Studies of Brain Metabolism in Psychiatric Patients: Can Standards Be Drawn?

Springer-Verlag   Wien New York

Dr. H. Ågren
Department of Psychiatry, University Hospital, Uppsala, Sweden

Dr. J.-L. Martinot
Department of Psychiatry, Hospital A. Chenevier, Créteil, France

Prof. Dr. F.-A. Wiesel
Department of Psychiatry, Ulleråker, University of Uppsala, Sweden

With 17 Figures

ISSN 0303-6995
ISBN-13:978-3-211-82346-0   e-ISBN-13:978-3-7091-9209-2
DOI: 10.1007/978-3-7091-9209-2

Preface

When, in the Seventies, Ingvar and Franzén found a decrease of the cerebral blood flow in the prefrontal regions of chronic schizophrenic patients, it became obvious that functional brain imaging techniques would contribute to the evaluation of brain regional activity during psychiatric illnesses. In addition, these techniques afforded the possibility to study directly what influence psychiatric disorders and their treatments might exert on the brain.

During the following years, several European and American research groups tried to replicate this finding in other schizophrenic patients, measuring cerebral regional glucose consumption with positron emission tomography (PET). It was confirmed by some groups, but not by all. The reasons put forth to explain these heterogeneous results concerned the diversity of PET cameras used, data analysis procedures, and the variety of diagnostic criteria and methods used to evaluate the symptomatology.

By 1990, the number of European groups involved in studies of psychiatric patients had dramatically increased (more than 10); most PET centers willing to study brain metabolism, but each tending to use its own methodology. This situation rendered it necessary to set up a minimum degree of standardization of the data, in order to preserve the "comparability" of information from different centers.

This special issue of the *Journal of Neural Transmission* presents the works by several European psychiatrists and scientists in the field of PET, gathered together in Orsay (France, December 1990), under the aegis of the "EEC Concerted Action on PET Investigations of Cellular Regeneration and Degeneration". It includes contributions focussing on the present states of research in this field, and on the questions and problems raised by the standardization of the measures in the clinical and therapeutical domains, but it also covers ligand modelling, data analysis, anatomical definition of the brain regions studied, and cerebral activation during experimentally controlled mental activities. This issue should be a useful source of information for all those interested in brain imaging.

H. ÅGREN
J.-L. MARTINOT
F.-A. WIESEL

Contents

Listed in Current Contents

J Neural Transm (1992) [Suppl] 37: 1–18

Glucose metabolism in psychiatric disorders: how can we facilitate comparisons among studies?

F.-A. Wiesel

Department of Psychiatry, Uppsala University, Uppsala, Sweden

Summary. Positron emission tomography (PET) offers a possibility to study brain function and its relationship to psychiatric disorders. Clinical studies have demonstrated that several psychiatric diseases are coupled with changes in brain glucose metabolism. Schizophrenia seems to involve a lower metabolism in wide areas of the brain — both cortical and subcortical structures. Depression probably involves dysfunction of the metabolism in dorsolateral prefrontal cortex. Obsessive compulsive disorder, panic disorder, anorexia nervosa and the experience of anxiety may involve increased metabolic rates. The results from the different studies do not allow quantitative comparisons or detailed analyses because of large differences in experimental and clinical methodology.

The term Good Clinical PET Practice (GCPP) is suggested to encourage standardization in clinical investigations. GCPP includes standardization of both experimental factors (lumped constant, arterialization, purity of tracer, regions of interest, relative rates) and clinical factors (state of the subject, wakefulness, anxiety, gender, course of the disease) in PET performance.

Introduction

During the last two decades several brain imaging techniques have been introduced in psychiatric research making it possible to study various aspects of brain function in relation to disease and treatment. The most powerful technique is positron emission tomography (PET). The technique has wide applications like the determination of energy consumption, blood flow, transport of substances across the blood brain barrier, receptor numbers and receptor distribution. The possibility to determine regional glucose metabolism has attracted great interest in psychiatric research due to its coupling with neuronal activity.

The first PET investigations of psychiatric patients were performed in the beginning of the eighties. The expectations of PET in psychiatry were high, but on the whole the results did not allow a simple elucidation of psychiatric diseases in relation to regional brain function. This fact has made some investigators to doubt the usefulness of PET in psychiatry (as well as in organic brain diseases). However, the most important question concerning the relationship between brain function and psychiatric disease has indeed been demonstrated by PET. Thus, schizophrenic patients with pronounced negative symptoms and considered to have "low psychic energy" do literally have a decreased brain metabolism. There are problems with inconsistant results but this seems to be explained by factors such as patient heterogeneity, small samples, how calculations of metabolic rates are performed, differences in resolution of cameras involving partial volume effects, and how to determine regions of interest. In the following, I will discuss methodological aspects, obtained results and offer suggestions for Good Clinical PET Practice (GCPP) in order to facilitate comparisons among studies. Some agreement on how to standardize investigations should be important for a healthy development of PET in clinical research.

Sources of variance in determining regional glucose metabolism

Models to calculate glucose metabolism

The positron camera collects radioactivity data after an intravenous injection of a radioactive metabolic tracer. In order to be functionally meaningful the data have to be processed following a theoretical model of the relevant biochemical process. Sokoloff and coworkers (cf. Sokoloff, 1985) have developed a method for calculating regional glucose metabolism by the use of radiolabelled 2-deoxyglucose (DG) (see the contribution by Wienhard, this volume). Glucose and 2-deoxyglucose are competitive substrates using the same carrier across the blood-brain barrier, and they are phosphorylated by the same enzyme (hexokinase) to glucose-6-phosphate and 2-deoxyglucose-6-phosphate (DG-6-P). However, unlike glucose-6-phosphate, DG-6-P is not further metabolized, but trapped in the tissue. DG can be labelled with a positron emitting isotope and injected i.v. It has been shown that the accumulated radioactivity adapted to a three compartment model (tracer in plasma and tissue, metabolized tracer in tissue) is equal to regional brain glucose metabolism (Phelps et al., 1979).

In this calculation, a constant, the "lumped constant", has to be used, which combines six other constants related to enzyme kinetics under distribution of glucose and deoxyglucose (Sokoloff, 1985). The value of the lumped constant will directly influence the measured metabolic level. Early determinations of the lumped constant gave a value of 0.42. However, later determinations in man have given values of 0.52 for ^{18}F-DG and 0.56 for ^{11}C-DG (Reivich et al., 1985). This means that the metabolic rate will be overestimated with 20–25 percent if the lower value of 0.42 is used for the lumped constant (see Wienhard, this volume). Unfortunately, it is not always clear which value of the lumped constant that has been used in clinical investigations.

^{11}C-glucose can also be used to determine brain glucose metabolism. The advantage in using ^{11}C-glucose as tracer is obvious: glucose is the natural substrate for brain sugar metabolism. However, radiolabelled glucose is not trapped in the tissue, but further metabolized, mainly to $^{11}CO_2$, which is a major drawback. This fact, and other aspects of using ^{11}C-glucose as the tracer, have been dealt with in a model developed by Blomqvist et al. (1990). Using ^{11}C glucose as tracer, brain glucose metabolism may be underestimated by 20 percent assuming that the lumped constant for ^{11}C-DG is correctly determined in man.

In calculating metabolic rates, the input function of the tracer into the brain is of great importance. This motivates the determination of tracer in arterial blood. However, since this is not always possible it has become a common practice to use arterialized venous blood from the hand. This appears to be quite accurate, at least if one uses ^{18}F-DG, but one should also determine the arterialization (pO_2) (see Wienhard, this volume). If radiolabelled glucose is used, arterial blood samples are necessary since even small differences in arterialization will influence the input function drastically, thereby influencing the estimation of the brain metabolism (Blomqvist et al., 1990; Wiesel et al., 1987).

Regions of interest

The use of PET cameras with different resolutions will complicate comparisons among studies, since the precision in determining regional metabolic rates is influenced by the resolution of the camera (full width at half maximum, FWHM). The resolution determines the minimum size of the objects that can be studied with the camera (Mazziota et al., 1981). In addition to the size of the region, the shape has also to be considered, both in the drawings and in the interpretation of the metabolic value. Small errors occur in large circular structures sourrended by regions of similar values. The so called *partial volume effect* is related to the structure, the size, and the shape of the region.

The early PET cameras had poor resolutions, to the degree that important questions related to psychiatric diseases could never be investigated. The new generation of cameras with resolutions of 3–5 mm and with thinner slices will make it possible to study small structures in the limbic system, parts of the thalamus and the brain stem, which might offer crucial information as for psychiatric disorders.

In determining metabolic rates the region of interest has to be correctly delineated. Using CT or MR images of the subjects own brain provides the most accurate way to define a region of interest. The defined regions of interest are then transferred to the corresponding PET image. However, this method is time-consuming, and the new generation of PET cameras with a high resolution and a slice thickness of 6–7 mm (generating 15 slices) makes the use of manual methods not so feasible.

There exists several automatic techniques that can be used for the outlining of regions of interest. The most sophisticated technique involves the use of "brain atlas" programs (see Bohm et al., this volume). Brain atlas program enables one to use all the information contained in the PET investigation, which is extremely important when applying various stimulations to a subject for the study of functional changes in relation to morphology (see Friston, this volume).

In psychiatric research it is pertinent to study inter-hemispheric functional differences. Using a high-resolution camera, left-right differences could easily be caused by small tiltings in one of the three planes. This can, at least to some extent, be taken into account with the new brain atlas programs. However, the reduction of this kind of variance necessitates that normals and patients are positioned in a reproducible way.

Variance factors among subjects

The state of the subject will influence the metabolism to be accounted for during the PET investigation. A "resting condition" implies a subject placed in a room with dampened noise and lights, with eyes either open or covered, and ears either plugged or unplugged. An early study demonstrated a progressive decline in overall glucose metabolism with reduced sensor inputs (Mazziota et al., 1982). A 75 percent higher metabolism was found in subjects stimulated visually compared with subjects with eyes closed as well as ears plugged. Obviously, this would indicate that pronounced differences among studies might be due to differences in resting conditions during investigations. On the other hand, in a recent study glucose metabolism was determined in two groups, one with eyes closed and one with eyes open. In the posterior calcarine cortex, metabolism declined by 14 percent with eye closure, otherwise no differences were found (Kiosawa et al., 1989).

Differences in wakefulness among subjects will also influence the metabolism. Thus during non-rapid eye movement sleep, a reduction in metabolism of about 23 percent was found across the entire brain (Buchsbaum et al., 1989). In our own investigations, wakefulness has been controlled by EEG recordings during the PET investigation (Wiesel et al., 1987).

One objection to investigations using a subject in the "resting condition" is the uncertainty as to what degree differences in unrestrained mental activity will influence the metabolism. Probably, only a small part of the variance can be explained by differences in mental activity during resting conditions. Indirect support for this view is provided by a study in which Bartlett et al. (1991) investigated the stability of the deoxyglucose metabolism in resting normal controls and in patients with schizophrenia. ^{11}C-DG was injected twice in a single day. The average change in the whole brain metabolism was 6 percent in the normals and 8 percent in the schizophrenics. These percentual changes are quite small when considering the complexity of the investigations. Most probably, the contribution of different mental activities to the variance should be minor.

Even with longer intervals between investigations, the stability of glucose consumption is striking. In one study the average variation between two investigations for all gray matter areas was $-8 \pm 15\%$ (Maquet et al., 1990). The overall coefficient of variance in whole brain metabolism varies by some 10–20 percent (Bartlett et al., 1991; Brooks et al., 1987; Duara et al., 1984; Maquet et al., 1990; Wiesel et al., 1987). This indicates that biological factors explain about half the variance. Well-controlled clinical studies should be able to detect 20 percent differences in metabolism between groups with rather small samples (n = 20).

The determination of subjects' handedness has almost become standard. To my knowledge there is no real evidence that resting metabolic rates differ between left- and right-handed individuals. In clinical studies most patient groups include both women and men. Baxter et al. (1987a) reported women's whole brain glucose metabolism to be 19 percent higher than that of men. However, another study whole brain glucose metabolism did not demonstrate significant sex differences, nor did any regional metabolic rate (Miura et al., 1990). One likely explanation for this difference might be related to the manner in which attenuation by brain and skull was corrected for. In average, females have smaller brains than males, and there may also be sex differences in skull thickness. Such differences are accounted for with a method of direct attenuation correction in each subject, as was done in the study by Imura and coworkers. The method used for attenuation correction is seldom reported in clinical studies.

Relative rates

Calculating normalized or relative values is one common way to reduce the influence of experimental factors on the variance. Relative rates are usually defined as the regional metabolic rates versus a defined denominator that could be the whole brain, the ipsilateral hemisphere, the whole brain slice or the ipsilateral brain slice. The use of relative rates increases the possibility to find differences between groups, since minor differences that may be obscured by some general arousal or depression of brain metabolism will be eliminated. Particularly the relative rates would facilitate comparisons across studies. A prerequisite for quantitative comparisions is then that the metabolic rate in the denominator must be identical among groups and studies. Otherwise, one will use different scales with different meanings of the percentage changes. Comparisons between studies are also complicated by the use of different denominators rendering quantitative comparisons less meaningful. In most cases one may only make qualitative comparisons. The need for standardization is demonstrated by the fact that whole brain metabolism varies twofold in the literature, from 20 to 40 µmol/100 g/min (cf. Wiesel, 1989). Standardization of PET procedures for clinical studies should make it possible to obtain large enough patient samples, allowing a more detailed analysis of brain function and psychiatric disease.

Clinical studies

Schizophrenia

Schizophrenia is the most extensively studied psychiatric disorder using PET. The results indicate that schizophrenic patients are characterized by reduced metabolic rates in both frontal and posterior cortical regions (Table 1). Some recent reviews have claimed that studies of glucose metabolism in schizophrenic patients have yielded inconsistant results. This criticism may be valid as for the concept of hypofrontality, but not for absolute metabolic rates. In fact, there are now several different studies demonstrating that patients with schizophrenia have a lower metabolism than normals in wide areas of the brain during a resting condition. However, acute patients suffering their first psychotic episode do not seem to differ in their energy consumption in comparison with normal controls (Cleghorn et al., 1989; Sheppard et al., 1983; Wiesel et al., 1987) Significant increases in metabolism have been reported when patients were given a painful electrical stimulus of the right forearm (Table 1).

Table 1. Reported changes in CMR_{Glc} in schizophrenic patients without neuroleptic treatment

---

Increases
Posterior cortex[1]
Temporal cortex left > right[2]

Decreases
Frontal, parietal, occipital and temporal lobes[3]
Frontal and temporoparietal cortex[4]
Whole cortex[5]
Frontal and parietal cortex[6]
Whole cortex[7]
Medial frontal and left temporal cortex[8]
Frontal, posterior and temporal cortex[9]
Frontal and posterior brain[10]
Whole cortex[11]
Basal ganglia areas[3,5,7,8,9]

---

[1] Buchsbaum et al., 1984 (stimulation); [2] DeLisi et al., 1989 (stimulation); [3] Bartlett et al., 1991 (from presented data, t tests were used to calculate differences between normals and patients); [4] Buchsbaum et al., 1990 (attention test); [5] Gur et al., 1987; [6] Huret et al., 1991; [7] Resnick et al., 1988; [8] Wiesel et al., 1987; [9,10] Wolkin et al., 1985, 1988; [11] Volkow et al., 1987

The causal relationship between glucose metabolism and schizophrenia is unknown. The observed changes in glucose metabolism in chronic patients may reflect an underlying brain dysfunction that is demasked after a period of time with overt disease. Schizophrenia is a set of complex syndromes involving changes in many aspects of brain function, but peripheral manifestations of the disease are also present. For example, neuromuscular abnormalities have been found in schizophrenic patients (Borg et al., 1987; Goode et al., 1977). In addition, tyrosine transport is decreased in fibroblasts from schizophrenic patients (Hagenfeldt et al., 1987), and, furthermore, a decreased influx rate of tyrosine across the blood-brain barrier as measured by PET has been demonstrated in a small group of schizophrenic patients (Wiesel et al., 1991). Altogether, the results may suggest that the metabolic changes observed in the brain of schizophrenics may be the consequence of a general deficit in cell membrane function.

The clinical symptomatology of schizophrenic patients is at least partly coupled with the observed changes in metabolic rates. Thus, autistic or negative symptoms were negatively correlated with metabolic rates (Wiesel et al., 1987; Wolkin et al., 1985). In accordance with these findings two other studies reported lower metabolic rates in patients with negative symptoms (Huret et al., 1991; Volkow et al., 1987). Increased metabolic rates have been coupled with high BPRS scores and high level of anxiety (Gur et al., 1987; Wik et al., 1991). Furthermore, one study has shown increased left temporal lobe glucose metabolism to be related to the severity of the disease (DeLisi et al., 1989).

Hypofrontality in schizophrenia is a widely discussed concept that stems from the findings of Ingvar and Franzén (1974). These authors described a lower blood flow in the frontal versus the posterior part of the brain (frontal versus parietal cortex) in schizophrenic patients, i.e. hypofrontality. Controls exhibited the opposite pattern. Hypofrontality, however, was clearly evident only in a group of older schizophrenic patients (mean age 61 years), not in younger patients. In young, never drug treated patients with schizophrenia, hypofrontality has never been reported (Sheppard et al., 1983; Wiesel et al., 1987; Cleghorn et al., 1989). The pattern found in a mixed group of acute and chronic patients (Wiesel et al., 1987) and in acute never treated patients (Cleghorn et al., 1989) is more consistant with being hypoparietal (Table 2). In a review of 17 studies hypofrontality was reported in 7 (from the reference list). This apparent inconsistency in results may be due to differences in the clinical state of the patients. Thus, a hypofrontal blood flow may be related to residual schizophrenia or to patients in remission (Geraud et al., 1987; Warkentin et al., 1990). Patients with acute exacerbations are more likely to have an anterior/posterior distribution of blood flow similar to normals.

Increased relative metabolic rates have been reported by several investigators in the area of the basal ganglia (Table 2). It cannot be determined whether this would indicate that subcortical structures are less affected in schizophrenic patients than are cortical areas, or whether it points to a

Table 2. Reported changes in relative CMR_{Glc} in schizophrenic patients without neuroleptic treatment

---

Increases
Frontal cortex[1]
Basal ganglia areas[2,3,4]

Decreases
Parietal cortex[1]
Medial frontal, left temporal and parietal cortex[3]
Frontal cortex[5]
Frontal and temporoparietal cortex[6]
Mid-prefrontal cortex[7]
Whole and frontal cortex[8]
Whole cortex[9]

---

[1] Cleghorn et al., 1989; [2] Resnick et al., 1988; [3] Wiesel et al., 1987; [4] Wolkin et al., 1985; [5,6] Buchsbaum et al., 1982, 1990 (attention test); [7] Cohen et al., 1989 (attention test); [8] Huret et al., 1991; [9] Kling et al., 1986

primary disturbance of the basal ganglia. Probably, the results from subcortical structures are likely to have been influenced by partial volume effects. A more detailed study of the basal ganglia and the limbic structures awaits a high-resolution camera.

Mood disorders

Baxter and coworkers (1985) found that subtypes of mood disorders had different metabolic rates. Bipolar depressed patients as compared with unipolar and manic bipolar patients had lower metabolic rates in most parts of the brain (Table 3). Martinot and coworkers (1990b) also found decreased cortical metabolism in depressed patients (seven bipolars, three unipolars). Buchsbaum and coworkers (1986) reported increased metabolic rates both in bipolar and in a small sample of unipolar patients (Table 3). However, these authors did not investigate their patients in a resting state; in fact, they received a painful electric stimulus of the right forearm. In a study by Post and coworkers (1987) the affectively ill patients (electrical stimulus) had significantly elevated left temporal lobe glucose utilization, but the majority of the patients had remitted from their depression which might have contributed to this result. Eight female patients with major depression and partaking in an attention test were found to have significantly lower left hemisphere activity than controls (Hagman et al., 1990). Metabolic rates were also lower in the basal ganglia and thalamus, especially in the left hemisphere.

Relative rates seem to give consistant results: the depressed state in both bipolar, unipolar and obsessive-compulsive disorders is coupled with a relatively lower metabolism in the dorsolateral prefrontal cortex (Table 4;

Table 3. Reported changes in CMR_{Glc} in untreated patients with mood disorders

Increases
Bipolar disorder depressed and unipolar depression: Frontal and posterior cortex[1]

Decreases
Bipolar disorder depressed: Whole brain[2]
Unipolar depression: Medial frontal cortex, basal ganglia, thalamus[3]
Major depression: Whole cortex[4]
Major depression: Whole brain[5]

---

[1] Buchsbaum et al., 1986 (stimulation); [2] Baxter et al., 1985; [3] Hagman et al., 1990 (attention test); [4] Martinot et al., 1990b; [5] Raichle et al., 1985 (oxygen metabolism)

Table 4. Reported changes in relative CMR_{Glu} in untreated patients with mood disorders

Increases not reported

Decreases
Unipolar depression: Caudate nucleus[1]
Bipolar disorder depressed and unipolar depression: Prefrontal cortex[2]
Bipolar disorder depressed: Hypofrontality right hemisphere[3]
Bipolar disorder depressed and unipolar depression: Mid-prefrontal cortex[4]
Unipolar depression: Lower left hemisphere activity[5]
Major depression: Frontal cortex[6]
Major depression: Right temporal cortex[7]

[1,2] Baxter et al., 1985, 1989; [3] Buchsbaum et al., 1986 (stimulation); [4] Cohen et al., 1989 (attention test); [5] Hagman et al., 1990 (attention test); [6] Martinot et al., 1990b; [7] Post et al., 1987 (stimulation)

Baxter et al., 1989). Similar findings have been reported by Martinot et al. (1990a), but with a more pronounced effect on the left side. Supporting these findings Buchsbaum and coworkers (1986) reported a significantly lower frontal-to-occipital cortex ratio in bipolar depressed patients. Relative decreases in metabolism have also been found in the basal ganglia of depressed patients (Baxter et al., 1985; Buchsbaum et al., 1986).

These results strongly indicate that the metabolic rate in the dorsolateral prefrontal cortex is affected in patients with depression. These changes, however, do not seem to discriminate between affective disorder and schizophrenia. Cohen and coworkers (1989) compared normals with schizophrenic and affectively disturbed patients during an attentional test. It was found that patients with affective disorder more often had lower relative metabolic rates in the temporal lobe and the left basal ganglia. Both groups had low relative metabolic rates in the mid-prefrontal cortex, higher rates in the superior parietal cortex, and lower rates in the hippocampal region of both patient groups. It was concluded that a brain dysfunction was

F.-A. Wiesel

Table 5. Reported changes in CMR_{Glc} in anxiety, eating disorders and alcohol dependence

Increases
Obsessive compulsive disorder: Orbital gyri, caudate nuclei, whole brain[1,2]; Frontal, sensorimotor and anterior cingulate cortex, thalamus[3]
Panic disorder: Whole brain[4]
Anorexia nervosa: Brainstem, caudate nuclei, thalamus[5]; Caudate nuclei[6]

Decreases
Obsessive compulsive disorder: Whole cortex, striatum, thalamus[7]
Bulimia nervosa: Medial frontal cortex[8]
Alcohol dependence: Several cortical areas and thalamus[9]
Acute intake of alcohol decreases CMR_{Glu} more in alcoholics than normals[10]

[1,2] Baxter et al., 1987b, 1989; [3] Swedo et al., 1989; [4] Reiman et al., 1986 (lactate sensitive); [5] Herholz et al., 1987; [6] Delvenne et al., 1990; [7] Martinot et al., 1990a; [8] Hagman et al., 1990 (attention test); [9] Wik et al., 1988; [10] Volkow et al., 1990

involved in both depression and schizophrenia, and that the determination of regional metabolism cannot be used diagnostically.

In order to explore and understand the functional meaning of the observed metabolic changes in depressed and schizophrenic patients other methods must be used, such as specific stimuli in the context of neuronal networks.

Anxiety disorders and obsessive compulsive disorder

In obsessive compulsive disorder the metabolism is increased in the orbital gyri and the caudate nuclei (Table 5; Baxter et al., 1987b). This finding was later confirmed by the same group indicating metabolic changes in the caudate and the orbital gyri to be relevant in this disorder (Baxter et al., 1988). The metabolic pattern indicated a disturbance in the same regions (Table 6). Swedo and coworkers (1989) reported increased metabolic rates to be prevalent in wide areas of the brain. Increased relative rates were especially found in the right prefrontal and the left anterior cingulate regions, in comparison with the controls. One other study again found increases in relative rates, indicating the frontal cortex and the basal ganglia to be important (Table 6; Benkelfat et al., 1990). However, in one study of patients with obsessive compulsive disorder without depression, decreases in the glucose metabolism were found in all cortical regions, the striatum and the thalamus (Martinot et al., 1990a). Furthermore, relative rates were decreased in the whole prefrontal lateral cortex. The reason for this discrepant result is unclear, but Martinot and coworkers point to some factors such as high metabolic rates in their controls, elderly patients with neuropsychological disturbances, and high levels of anxiety in Baxter's patients.

Table 6. Reported changes in relative CMR_{Glc} in anxiety, eating disorders and alcohol dependence

---

Increases

Obsessive compulsive disorder: Orbital gyri[1,2]; Right prefrontal and left anterior cimgulate cortex[3]; Orbital frontal cortex, left caudate nucleus, left putamen[4]
Panic disorder: Right parahippocampus[5]
Anorexia nervosa: Caudate nuclei[6]

Decreases

Obsessive compulsive disorder: Left prefrontal cortex if depressed[7]; Prefrontal cortex[8]
Bulimia nervosa: Lower right hemisphere activity[9]
Alcohol dependence: Medial prefrontal cortex[10]; Parietal cortex[11]

[1,2] Baxter et al., 1987b, 1988; [3] Swedo et al., 1989; [4] Benkelfat et al., 1990; [5] Reiman et al., 1986 (lactate sensitive); [6] Krieg et al., 1991; [7] Baxter et al., 1989; [8] Martinot et al., 1990a; [9] Wu et al., 1990 (attention test); [10] Samson et al., 1986; [11] Wik et al., 1988

On the whole, metabolic PET results indicate a functional disturbance of the frontal-limbic-basal ganglia system which is in line with findings from psychosurgery studies (Mindus, 1991).

The idea that brain energy metabolism is increased in a state of anxiety was first presented by Kety (1950) who found increased oxygen consumption in the brain of one subject during a state of "grave apprehensiveness". Reiman and coworkers (1986) have also found an increased oxygen consumption in patients with panic disorder sensitive to lactate infusion. These patients had also lower left to right ratios of parahippocampal blood flow and oxygen consumption. In a later study the same group found that anticipatory anxiety in normals were coupled with blood flow increases bilaterally in the temporal poles (Reiman et al., 1989). Increased glucose metabolism in anxiety has also been found in normals in relation to dreams and in immediate relation to the PET investigation (Gottschalk et al., 1991; Wik and Wiesel, 1991). However, other PET studies of brain glucose metabolism and anxiety in normals did not observe any positive relationship between anxiety and metabolism. Even a curvilinear relationship has been found between anxiety and frontocortical metabolic rates (Reivich et al., 1984). Later that same group reported a linear decrease in metabolism with increased anxiety (Gur et al., 1984). Gioardiani et al. (1990) did not observe any relationships at all between anxiety and metabolism. In the study by Wik and Wiesel (1991) ^{11}C-glucose was used as tracer. A positive relationship between the intensity of anxiety during the investigation and the metabolism was seen in both normals and patients with schizophrenia. Positive correlations were found in most regions.

Analyses of relative rates indicated that the right middle frontal cortex and the left thalamus may be inhibited in a state of anxiety. The discrepant results may be due to experimental differences. Our subjects were exposed to stress in immediate relation to the tracer injection — the helmet was put on and the subject's head position was fixed in the camera just the minutes before ^{11}C glucose was injected and the scanning was started. Metabolic rates were calculated from data collected during the first 15 minutes. In the other studies referred above the subjects probably waited under relaxed conditions for a varied length of time before ^{18}F-DG was injected. The tracer was then incorporated during a period of 40 minutes in a dimly lit room with ambient noise reduced to a minimum before the scanning was performed. It seems uncertain if ^{18}F-DG under these conditions is a useful tracer in relating anxiety to glucose metabolism.

Anxiety in panic disorder and anxiety during anticipatory conditions seem to involve increased metabolism and blood flow. However, patients with generalised anxiety disorder had lower absolute metabolic rates in the basal ganglia and white matter during a passive viewing task (Wu et al., 1991). It is uncertain whether the patients experienced anxiety after the injection of ^{18}F-DG. Relative rates were higher in parts of the occipital, temporal, and frontal lobes and the cerebellum in relation to normals. The results are somewhat different and it is not possible to decide if this is due to differences in diagnosis or in experimental conditions.

Eating disorders

In a small sample of female patients with anorexia nervosa an increased metabolism was found in the caudate nucleus bilaterally (Herholz et al., 1987). Similar findings were reported by another group showing a relative increase in both caudate nuclei (Delvenne et al., 1990). Female patients with bulimia did not show these changes (Krieg et al., 1991). However bulimia may involve less right lateralization of both cortical and subcortical structures in comparison with normals (Wu et al., 1990).

Alcoholism

Regional brain glucose metabolism has been studied in a small group of male alcohol-dependent patients (Wik et al., 1988). The patients were socially impaired by the abuse and abstinent from alcohol and drugs for more than four weeks before entering the study. The alcoholics had 20–30 percent lower glucose metabolism than the controls in both cortical and subcortical regions. Relative rates indicated that parietal cortical areas were most affected. In another patient study no changes in metabolism were found but a relative decrease in the metabolism was observed in the medio-frontal cortex (Samson et al., 1986). The discrepancy in results between the studies was probably due to differences in the patient material and how regions of interest were delineated. Wik's patients had a more serious sequelae of their alcohol dependence. In another study acute administration of ethanol was found to inhibit cortical and cerebellar glucose metabolism and to a lesser degree basal ganglia metabolism (Volkow et al., 1990). This inhibitition was more pronounced in alcoholics than in controls. The results are in line with Wik's results indicating that protracted use of alcohol will cause longstanding changes in brain function.

Discussion

Results from different patient categories demonstrate that psychiatric diseases involve changes in brain metabolism. This statement is of particular importance to psychiatry. However, a detailed analysis of the relationship between brain function and psychiatric disease has not been possible. The experience of reviewing the clinical studies calls for standardization of the PET investigations and the term Good Clinical PET Practice is suggested (see also Wienhard, this volume). This may offer a possibility to combine results from different studies to obtain higher patient numbers which is very difficult for the individual PET investigator. Today only qualitative comparisons can be made, not quantitative.

Different lumped constants are used; sometimes the constant used is not given. Arterialized blood is frequently used, but the degree of arterialization is not often determined. It is uncertain if purity of the tracer is controlled. In the studies reviewed there are substantial differences how regions of interest were determined. Problems of partial volume effects must be substantial in the low resolution cameras. The importance of the conditions during the PET investigation is illustrated by studies of Buchsbaum who first used a painful electrical stimulus of the right forearm and with that technique found increases in some regions in patients with schizophrenia, but later when using a visual attention test he obtained results similar to other investigators who had the patient in a resting condition. It is still uncertain how sex and handedness of the subjects influence metabolic rates.

Chronic course of a disease is more likely to be linked with reduced metabolism than a disease in its early phase. Impaired cognitive function may also be related to a lower metabolism. Relative rates could not be used for quantitative comparisons and compilation of data, since different denominators were used and there were too big differences in metabolic levels among studies. With these problems in mind when comparing psychiatric studies, however, it seems still possible to claim that chronic schizophrenia involves brain dysfunction, but the dysfunction cannot be localized to any specific region. Reduction in metabolism seems to be coupled with negative symptoms. Mood disorders seem to especially involve a dysfunction of the prefrontal cortex. Increased metabolic rates seem to be related to obsessive compulsive disorder (due to anxiety?), panic disorder if sensitive to lactate. Patients with generalized anxiety disorder may have decreases in subcortical metabolism and changes in parts of the neocortex.

It is probable that the experience of intense anxiety is related to an increased metabolism.

In order to facilitate comparisons and for new researchers in the field one may attempt to give guidelines for Good Clinical PET Practice as follows. Use the lumped constant according to Reivich et al. (1985). If this is not possible, state which constant is used. If arterialized blood is used — determine the degree of arterialization. The chemical purity of the tracer used should be controlled. The subjects brain morphology should be investigated with MR or CT, and this information should be the basis for using a Brain Atlas. The error of the fixation or corresponding systems should be determined. The size of regions of interest should be in line with the resolution of the camera. If one uses a resting condition one must check that the subject is awake. Use average hemisphere activity as the denominator in the calculation of relative· rates. Try to determine the importance of different baseline clinical characteristics for the outcome of the investigation. If heterogenous patient materials are used, this must be taken into account in the statistical anlysis.

Good Clinical PET Practice may be achieved in many ways, but the suggestions offered may be considered as a starting point for discussion.

Acknowledgements

Ms. A. Liberg is gratefully acknowledged for preparing the manuscript. The study was supported by the Swedish Medical Council, grant 8318, and by the E.E.C. Concerted Action on PET Investigations.

References

Bartlett EJ, Barouche F, Brodie JD, Wolkin A, Angrist B, Rotrosen J, Wolf AP (1991) Stability of resting deoxyglucose metabolic values in PET studies of schizophrenia. Psychiatry Res Neuroimaging 40: 11–20

Baxter LR, Phelps ME, Mazziotta JC, Schwartz JM, Gerner RH, Selin CE, Sumida RM (1985) Cerebral metabolic rates for glucose in mood disorders. Arch Gen Psychiatry 42: 441–447

Baxter LR Jr, Mazziotta JC, Phelps ME, Selin CE, Guze BH, Fairbanks L (1987a) Cerebral glucose metabolic rates in normal human females versus normal males. Psychiatry Res 21: 237–245

Baxter LR, Phelps ME, Mazziotta JC, Guze BH, Schwartz JM, Selin CE (1987b) Local cerebral glucose metabolic rates in obsessive-compulsive disorder. Arch Gen Psychiatry 44: 211–219

Baxter LR, Schwartz JM, Mazziotta JC, Phelps ME, Pahl JJ, Guze BH, Fairbanks L (1988) Cerebral glucose metabolic rates in nondepressed patients with obsessive-compulsive disorder. Am J Psychiatry 145: 1560–1563

Baxter LR, Schwartz JM, Phelps ME, Mazziotta JC, Guze BH, Selin CE, Gerner RH, Sumida RM (1989) Reduction of prefrontal cortex glucose metabolism common to three types of depression. Arch Gen Psychiatry 46: 243–250

Benkelfat C, Nordahl TD, Semple WE, King C, Murphy DL, Cohen RM (1990) Local cerebral glucose metabolic rates in obsessive-compulsive disorder. Arch Gen Psychiatry 47: 840–848

Blomqvuist G, Stone-Elander S, Halldin C, Roland PE, Widén L, Lindqvist M, Swahn C-G, Långström B, Wiesel F-A (1990) Positron emission tomographic measurements of cerebral glucose utilization using [1-^{11}C]D-glucose. J Cereb Blood Flow Metab 10: 467–483

Borg J, Edström L, Bjerkenstedt L, Wiesel FA, Farde L, Hagenfeldt L (1987) Muscle biopsy findings, conduction velocity and refractory period of single motor nerve fibres in schizophrenia. J Neurol Neurosurg Psychiatry 50: 1655–1664

Brooks RA, Hatazawa J, Di Chiro G, Larson SM, Fishbein DS (1987) Human cerebral glucose metabolism determined by positron emission tomography: a revisit. J Cereb Blood Flow Metab 7: 427–432

Buchsbaum MS, Ingvar DH, Kessler R, Waters RN, Cappeletti J, van Kammen DP, King C, Johnson JL, Manning RG, Flynn RW, Mann LS, Bunney WE Jr, Sokoloff L (1982) Cerebral glucography with positron tomography. Arch Gen Psychiatry 39: 251–259

Buchsbaum MS, Delisi LE, Holcomb HH, Cappelletti J, King AC, Johnson J, Hazlett E, Dowling-Zimmerman S, Post RM, Morihisa J, Carpenterr W, Cohen R, Pickar D, Weinberger DR, Margolin R, Kessler RM (1984) Anteroposterior gradients in cerebral glucose use in schizophrenia and affective disorders. Arch Gen Psychiatry 41: 1159–1166

Buchsbaum MS, Wu J, Delisi LE, Holcomb H, Kessler R, Johnson J, King AC, Hazlett E, Langston K, Post RM (1986) Frontal cortex and basal ganglia metabolic rates assessed by positron emission tomography with [^{18}F]2-deoxyglucose in affective illness. J Affect Dis 10: 137–152

Buchsbaum MS, Gillin JC, Wu J, Hazlett E, Sicotte N, Dupont RM, Bunney WE Jr (1989) Regional cerebral glucose metabolic rate in human sleep assessed by positron emission tomography. Life Sci 45: 1349–1356

Buchsbaum MS, Nuechterlein KH, Haier RJ, Wu J, Sicotte N, Hazlett E, Asarnow R, Potkin S, Gich S (1990) Glucose metabolic rate in normals and schizophrenics during the continuous performance test assessed by positron emission tomography. Br J Psychiatry 156: 216–227

Cleghorn JM, Garnett ES, Nahmias C, Firnau G, Brown GM, Kaplan R, Szechtman H, Szechtman B (1989) Increased frontal and reduced parietal glucose metabolism in acute untreated schizophrenia. Psychiatry Res 28: 119–133

Cohen RM, Semple WE, Gross M, Nordahl TD, King AC, Pickar D, Post RM (1989) Evidence for common alterations in cerebral glucose metabolism in major affective disorders and schizophrenia. Neuropsychopharmacology 2: 241–254

Delvenne V, Lotstra F, Goldman S, Mendelbaum K, Appelbaum-Fondu J, Bidaut LM, Luxen A, Schoutens A, Mendlewicz J (1990) Caudate hypermetabolism in eating disorders detected by ^{18}F-fluorodeoxyglucose method and positron emission tomography. In: 17th Congress of Collegium Internationale Neuro-Psychopharmacologicum, Kyoto, Japan, p 330

DeLisi LE, Buchsbaum MS, Holcomb HH, Langston KC, King AC, Kessler R, Pickar D, Carpenter T Jr, Morihisa JM, Margolin R, Weinberger RD (1989) Increased temporal lobe glucose use in chronic schizophrenic patients. Biol Psychiatry 25: 835–851

Duara R, Grady C, Haxby J, Ingvar D, Sokoloff L, Margolin RA, Manning RG, Cutler NR, Rapoport SI (1984) Human brain glucose utilization and cognitive function in relation to age. Ann Neurol 16: 702–713

Geraud G, Arné-Bes C, Guell A, Bes A (1987) Reversibility of hemodynamic hypofrontality in schizophrenia. J Cereb Blood Flow Metab 7: 9–12

Giordani B, Boivin MJ, Berent S, Betley AT, Koeppe RA, Rothley JM, Modell JG, Hichwa RD, Kuhl DE (1990) Anxiety and cerebral cortical metabolism in normal persons. Psychiatry Res Neuroimaging 35: 49–60

Goode DJ, Meltzer HY, Caryton JW, Mazura TA (1977) Physiologic abnormalities of the neuronmuscular system in schizophrenia. Schizophr Bull 3: 121–138

Gottschalk LA, Buchsbaum MS, Gillin JC, Wu JC, Reynolds CA, Herrera DB (1991) Anxiety levels in dreams: relation to localized cerebral glucose metabolic rate. Brain Res 538: 107–110

Gur CR, Gur RE, Resnick SM, Skolnick BE, Alavi A, Reivich M (1984) The effect of anxiety on cortical cerebral blood flow and metabolism. J Cereb Blood Flow Metab 7: 173–177

Gur RE, Resnick SM, Alavi A, Gur RC, Caroff S, Dann R, Silver FL, Saykin AJ, Chawluk JB, Kushner M, Reivich M (1987) Regional brain function in schizophrenia. I. A positron emission tomography study. Arch Gen Psychiatry 44: 119–125

Hagenfeldt L, Venizelos N, Bjerkenstedt L, Wiesel FA (1987) Decreased tyrosine transport in fibroblasts from schizophrenic patients. Life Sci 41: 2749–2757

Hagman JO, Buchsbaum MS, Wu JC, Rao SJ, Reynolds CA, Blinder BJ (1990) Comparison of regional brain metabolism in bulimia nervosa and affective disorder assessed with positron emission tomography. J Affect Disord 19: 153–162

Herholz K, Krieg JC, Emrich HM, Pawlik G, Beil C, Pirke KM, Pahl JJ, Wagner R, Wienhard K, Ploog D, Heiss W-D (1987) Regional cerebral glucose metabolism in anorexia nervosa measured by positron emission tomography. Biol Psychiatry 22: 43–51

Huret JD, Mazoyer BM, Lesur A, Martinot JL, Pappata S, Baron JC, Lemperiere T, Syrota A (1991) Cortical metabolic patterns in schizophrenia: a mismatch with the positive-negative paradigm. Eur Psychiatry 6: 7–19

Ingvar DH, Franzen G (1974) Abnormalities of cerebral blood flow distribution in patients with chronic schizophrenia. Acta Psychiatr Scand 50: 425–462

Kety SS (1950) Circulation and metabolism in the human brain in health and disease. Am J Med 8: 205–217

Kiosawa M, Bosley TM, Kushner M, Jamieson D, Alavi A, Savino PJ, Sergott RC, Reivich M (1989) Positron emission tomography to study the effect of eye closure and optic nerve damage on human cerebral glucose metabolism. Am J Ophthalmol 108: 147–152

Kling AS, Metter EJ, Riege WH, Kuhl DE (1986) Comparison of PET measurement of local brain glucose metabolism and CAT measurement of brain atrophy in chronic schizophrenia and depression. Am J Psychiatry 143: 175–180

Krieg J-C, Holthoff V, Schrieber W, Pirke KM, Herholz K (1991) Glucose metabolism in the caudate nuclei of patients with eating disorders, measured by PET. Eur Arch Psychiatry Clin Neurosci 240: 331–333

Maquet P, Dive D, Salmon E, von Frenckel R, Franck G (1990) Reproducibility of cerebral glucose utilization measured by PET and the [^{18}F]2-fluoro-2-deoxy-D-glucose method in resting, healthy human subjects. Eur J Nucl Med 16: 267–273

Martinot JL, Allilaire JF, Mazoyer BM, Hantouche E, Huret JD, Legaut-Demare F, Deslauriers AG, Hardy P, Pappata S, Baron JC, Syrota A (1990a) Obsessive-compulsive disorder: a clinical neuropsychological and positron emission tomography study. Acta Psychiatr Scand 82: 233–242

Martinot JL, Hardy P, Feline A, Hure JD, Mazoyer B, Attar-Levy D, Pappata S, Syrota A (1990b) Left prefrontal glucose hypometabolism in the depressed state: a confirmation. Am J Psychiatry 147: 1313–1317

Mazziotta JC, Phelps ME, Plummer D, Kuhl DE (1981) Quantitation in positron emission computed tomography. 5. Physical-anatomical effects. J Comput Assist Tomogr 5: 734–743

Mazziotta JC, Phelps ME, Carson RE, Kuhl DE (1982) Tomographic mapping of human cerebral metabolism: sensory deprivation. Ann Neurol 12: 435–444

Mindus P (1991) Capsulotomy in anxiety disorders. A multidisciplinary study. Thesis, Karolinska Institutet, Stockholm

Miura SA, Schapiro MB, Grady CL, Kumar A, Salerno JA, Kozachuk WE, Wagner E, Rapoport SI, Horwitz B (1990) Effect of gender on glucose utilization rates in healthy humans: a positron emission tomography study. J Neurosci Res 27: 500–504

Phelps ME, Huang SC, Hoffman EJ, Selin C, Sokoloff L, Kuhl DE (1979) Tomographic measurement of local cerebral glucose (F-18)2-fluoro-2-deoxy-D-glucose: validation of method. Ann Neurol 6: 371–388

Post RM, DeLisi LE, Holcomb HH, Uhde TW, Cohen R, Buchsbaum MS (1987) Glucose utilization in the temporal cortex of affectively ill patients: positron emission tomography. Biol Psychiatry 22: 545–553

Raichle ME, Taylor JR, Herscovitch P, Guze SB (1985) Brain circulation and metabolism in depression. In: Greitz T, Ingvar DH, Widén L (eds) The metabolism of the human brain studied with positron emission tomography. Raven Press, New York, pp 453–456

Reiman EM, Raichle ME, Robins E, Butler FK, Herscovitch P, Fox P, Perlmutter J (1986) The application of positron emission tomography to the study of panic disorder. Am J Psychiatry 143: 469–477

Reiman EM, Fusselman MJ, Fox PT, Raichle ME (1989) Neuroanatomical correlates of anticipatory anxiety. Science 243: 1071–1074

Reivich M, Alavi A, Gur RC (1984) Positron emission tomographic studies of perceptual tasks. Ann Neurol [Suppl] 15: 61–65

Reivich M, Alavi A, Wolf A, Fowler J, Russell J, Arnett C, MacGregor RR, Shiue CY, Atkins H, Anand A, Dann R, Greenberg JH (1985) Glucose metabolic rate kinetic model parameter determination in humans: the lumped constants and rate constants for [^{11}C]deoxyglucose. J Cereb Blood Flow Metab 5: 179–192

Resnick SM, Gur RE, Alavi A, Gur RC, Reivich M (1988) Positron emission tomography and subcortical glucose metabolism in schizophrenia. Psychiatry Res 24: 1–11

Samson Y, Baron JC, Feline A, Bories J, Crouzel C (1986) Local cerebral glucose utilisation in chronic alcoholics: a positron tomographic study. J Neurol Neurosurg Psychiatry 49: 1165–1170

Sheppard G, Manchanda R, Gruzelier J, Hirsch SR (1983) ^{15}O positron emission tomographic scanning in predominantly never-treated acute schizophrenic patients. Lancet ii: 1448–1452

Sokoloff L (1985) Basic principles in imaging of regional cerebral metabolic rates. In: Sokoloff L (ed) Brain imaging and brain function. Raven Press, New York, pp 21–49

Swedo SE, Schapiro MB, Grady CL, Cheslow DL, Leonard HL, Kumar A, Friedland R, Rapoport SL, Rapoport JL (1989) Cerebral glucose metabolism in childhood-onset obsessive-compulsive disorder. Arch Gen Psychiatry 46: 518–523

Volkow ND, Wolf AP, Van Gelder P, Brodie JD, Overall JE, Cancro R, Gomez-Mont F (1987) Phenomenological correlates of metabolic activity in 18 patients with chronic schizophrenia. Am J Psychiatry 144: 141–158

Volkow ND, Hitzeman R, Wolf AP, Logan J, Fowler JS, Christman D, Dewey SL, Schyler D, Burr G, Vitkun S, Hirschowitz J (1990) Acute effects of ethanol on regional brain glucose metabolism and transport. Psychiatry Res Neuroimaging 35: 39–48

Warkentin S, Nilsson A, Risberg J, Karlson S, Flekköy K, Franzén G, Gustafson L, Rodriguez G (1990) Regional cerebral blood flow in schizophrenia: repeated studies during a psychotic episode. Psychiatry Res Neuroimaging 35: 27–38

Wiesel F-A (1989) Positron emission tomography in psychiatry. Psychiat Dev 1: 19–47

Wiesel F-A, Wik G, Sjögren I, Blomqvist G, Greitz T, Stone-Elander S (1987) Regional brain glucose metabolism in drug-free schizophrenic patients and clinical correlates. Acta Psychiatr Scand 76: 628–641

Wiesel FA, Blomqvist G, Halldin C, Sjögren I, Bjerkenstedt L, Venizelos N, Hagenfeldt L (1991) The transport of tyrosine into the human brain as determined with L-[1-^{11}C]tyrosine and PET. J Nucl Med 32: 2043–2049

Wik G, Wiesel F-A (1991) Regional brain glucose metabolism and correlations to biochemical measures and anxiety in patients with schizophrenia. Psychiatry Res Neuroimaging 40: 101–114

Wik G, Borg S, Sjögren I, Wiesel F-A, Blomqvist G, Borg J, Greitz T, Nybäck H, Sedvall G, Stone-Elander S, Widén L (1988) PET determination of regional cerebral glucose metabolism in alcohol-dependent men and healthy controls using ^{11}C-glucose. Acta Psychiatr Scand 78: 234–241

Wolkin A, Jaeger J, Brodie JD, Wolf AP, Fowler J, Rotrosen J, Gomez-Mont F, Cancro R (1985) Persistence of cerebral metabolic abnormalities in chronic schizophrenia as determined by positron emission tomography. Am J Psychiatry 142: 564–571

Wolkin A, Angrist B, Wolf A, Brodie JD, Wolkon RN, Jaeger J, Cancro R, Rotrosen J (1988) Low frontal glucose utilization in chronic schizophrenia: a replication study. Am J Psychiatry 145: 251–253

Wu JC, Hagman J, Buchsbaum MS, Blinder B, Derrfler M, Tai WT, Hazlett E, Sicotte N (1990) Greater left cerebral hemispheric metabolism in bulimia assessed by positron emission tomography. Am J Psychiatry 147: 309–312

Wu JC, Buchsbaum MS, Hershey TG, Hazlett E, Sicotte N, Johnson JC (1991) PET in generalized anxiety disorder. Biol Psychiatry 29: 1181–1199

Author's address: Dr. F.-A. Wiesel, Department of Psychiatry, Uppsala University, S-750 17 Uppsala, Sweden

J Neural Transm (1992) [Suppl] 37: 19–25

Operators and scales: diagnostic and rating issues in psychiatric PET research

H. Ågren

Department of Psychiatry, University Hospital, Uppsala, Sweden

Summary. In psychiatric research that for various reasons has to restrict itself to a limited number of subjects, such as studies involving expensive positron emission tomography techniques, issues concerning the parsimonious description of patients gain in importance. The number of descriptive variables must be optimally small. This paper offers a conceptual back-ground for the choice of operators in operational diagnostic systems designed to delimit pathological types, and of rating scales designed to measure syndromal severity in a dimensional way. A practical suggestion in five tenets for the organization of clinical research of this kind is presented.

Introduction

The costs of PET experiments raise practical obstacles against accumulating large patient samples. We are presented with an awkward situation with a large number of potentially meaningful pieces of clinical information, and an even larger volume of information from each PET experiment, but few subjects on whom to run the correlations. The scarcity of subjects will restrict testing of only very strong hypotheses by correlating the clinic with the PET images. The dangers of committing Type II errors by exercising "too many" correlations without a proper post hoc adjustment of the standard significance level may lurk at many levels, and can be treacherous in data-exploring multivariate analyses.

Nosological specificity of neurobiological mechanisms has often been discussed but solid findings remain rare. Correlational results between clinical data and psychobiological variables have typically been successful with diagnosis-transgressing symptoms/syndromes — the serotonin-aggression story is illustrative, indices of deficient brain serotonin having been linked with suicidality and/or impulsivity in unipolar depression as well as in personality disorders (reviewed by Coccaro, 1989). Superficially, it would be easy to use multivariate statistical analyses to determine a variety of independent biological variables predicting one categorical clinical variable (y), in simple models:

$$y = x_1 + x_2 + \ldots + x_n$$

The difficult part lies in interpretation of significant partial coefficients and interactions among "independent" predictors. Rules about grouping of independent variables into "sets" are discussed in a great variety of texts on multivariate statistics and multiple regression, demonstrating several ways to deal with spurious significances. In small n studies, where the number of interesting x predictors may even exceed the number of observations, regressions may become quite meaningless. Thus, the task is to relate the vast amount of potentially useful information from PET images to a potentially equally vast volume of clinical information (with the multitude of symptom items).

A study involving a larger number of observations (large n study) would somehow allow interpretation of the many possibly relevant correlations reaching statistical significance. However, in the practice of PET research the small n will always be a "rate-limiting step" for meaningful statistical analyses. As a remedy, the number of research variables should be kept small, both for the PET calculations and for the clinical ratings. On the clinical side, this will allow comparisons with clinical diagnostic and sub-diagnostic entities, with global rating scale scores and perhaps with subscale scores, but *not* with the large array of symptom items that build up the scales (unless there is a very good a priori reason for it). Demonstrated reproducibility of an unexpected linkage between some symptoms and a biologic variable may, of course, turn out to be a truly serendipitous finding of great importance, but this situation is rare.

A thoughtful use of multivariate statistical procedures such as principal component analysis, helping reduce the number of variables, might be advocated under certain circumstances. It is important to point out that in order to reduce the incidence of false results, the clinical meanings of the deduced components have to be grasped intuitively by the experienced clinician. A few potentially meaningful components might then be compared with PET data.

It is not feasible to state a minimal number of patients that should be included in a statistically sound PET study. By the nature of the measured PET variables, they are often given together with standard errors (for example slopes), which might even justify the comparison of one measure on one individual and the same measure on another, or the same measure evaluated twice on the same individual that was followed longitudinally. In the case of slopes, a simple analysis of covariance would reveal any statistically significant difference between the slopes themselves (significant interaction between the categorical x variable and a continuous x variable) or in the intercepts with the y axis (a significant categorical x variable). However, if standard errors cannot be deduced, the numbers behave like any ordinary biological measurement of a level, and statistically significant differences can only be detected if the number of individuals inestigated is "large enough". The issue of statistical power can at its most basic level be formulated thus: a probability level of less than 5 percent is always interesting, even if $n = 5$–10 in both groups compared, provided the a priori hypothesis to find such a difference was a good one.

Typology vs. dimensionality

Medical disease entities are *types*, whereas psychological scales are *dimensions*. The use of either one of these, or both, in psychiatric research was the subject of an emotional debate between psychiatrists and psychologists in the 70'ies. The one-sided stand has more or less disappeared from the scene in the past few years. Epistemologically, borders between disease entities or types represent hypotheses about their proper grouping (Roth, 1978). Their purpose is to better envisage some "hidden" pathological structure or process which is surmised to be a *type*. Hypotheses on the structure of typological diagnoses evolve continuously into more refined concepts or are, at rare instances, replaced by entirely new logical cutting ideas.

Ever since Thomas Sydenham founded modern clinical nosology over 300 years ago, the rules of proper diagnostic decision flow in medicine has stated that groups of *covarying symptoms and signs* form *syndromes*, whereupon information about longitudinal time course is introduced in the emergent concept of *disease* or *illness*, which should if possible, but not necessarily, carry an implicit connotation of known etiology or pathogenesis. The term *disorder* has been popularized by recent American usage (Klein, 1978), and seems to stand inbetween syndrome and disease as for implied knowledge of chronology. Operational criteria have been created to keep track in this process ever since the first set of formalized criteria were published twenty years ago (Feighner et al., 1972). These criteria formed the basis of the Research Diagnosis Criteria (RDC; see Spitzer et al., 1978) and the subsequent DSM-III process.

The concept "operator" can be defined as a *"tool" agreed upon for describing a non-unique characteristic of an individual with a hidden disease type (the "hidden structure")*. The tool can be of any conceptual kind — symptoms and signs and temporal information being the most common — and together they may present such a very mixed bag of characteristics that the somewhat pejorative term "Chinese smörgåsbord" has been used. Operators are selected from merits such as communicability, good interrater reliability and mutual non-overlap, and they emphatically do *not* describe the full disease picture. Symptom operators are individually non-necessary rules, since there exists no single operator that must be necessarily present to warrant the diagnosis of a specific illness. Temporal operators, however, can be necessary — for example the 6 month rule of continuous symptoms for the diagnosis of DSM-III schizophrenia.

Clinical nosographical operators can be compared to biological tests. Both organize knowledge from one selected vantage point, which might well be termed a *window* into the brain. Commonly used windows are:

Psychoendocrine window . . . . *Tests of anterior pituitary function*

Electrophysiological window . . . . *Records of cortical potentials*

Neuropsychological window . . . . *Cognitive and motor tests*

"Mental Symptoms and Signs window" . . *Present mental state*

Operators and/or scales?

Typological diagnostic systems using *operators* (such as RDC, DSM-III or DSM-III-R) must be carefully distinguished from the multitide of *scales* that have been constructed to measure severity of isolated symptoms or of syndromes (BPRS, MADRS, etc.; see below).

The use of *operators* assume a hidden disease type, i.e. the presence of a clinical diagnostic hypothesis on the existence of an illness.

Superficially, the *exclusive use of scales* would appear to be suitable to allow an unbiassed clinical evaluation. However, this does not permit analysis of the interplay between observations and the launching of hypotheses which lies at the core of scientific thinking, and it cannot promote an understanding of any systems disorder in the CNS. Scales measure the loudness of voices of individual actors on the scene, but the relatedness of their cues is not evident.

The *exclusive use of operators* would make it impossible to distinguish between mild or severe forms of the same illness. This distinction is necessary when using *time* as a research variable in a longitudinal set-up. However, the distinction is unnecessary when doing simple comparisons between a pathological group and a healthy group of individuals.

In sum, a well-designed use of operators is of primary concern in psychiatric research. Scales are necessary in longitudinal set-ups and for confirmation of typology-based deviations. They may also be useful in situations when no true healthy group is available for comparison, so that patients with low scores form a counter-group ("healthier") against the very sick.

How should psychiatric patients be characterized for purposes of PET research? A five-folded suggestion

Grouping of predefined types

Groups have to be defined operationally and shown to be *types* according to RDC, DSM-III, DSM-III-R, *or one's own educated definition.* There is no good reason to avoid the recent DSM-III-R. Comparisons between a small number of separate types (one of which may be healthy volunteers) is a traditional and economic way to compare findings in limited sample sizes of test subjects.

Redefinition of types as need arises

If an interesting but still somewhat equivocal finding emerges one can modify inclusion criteria into groups and redo the comparison. The diagnostic operators/criteria should be kept on file so that changes can be done expediently. After all, biological findings are to be expected to modify our clinical diagnostic habits, even though no such rules have as yet gained common clinical acceptance.

Dimensions and scales

Correlations with severity measures (dimensions) may support the inductive reasoning in the two preceding steps. One should primarily only use total scale scores or a small number of subscale scores (or principal components). Correlations with individual items can be seductive and are defensible only if they are suggested by a preformed hypothesis. Uncritical correlations with every possible item will by statistical necessity produce some spurious significant results (problem of "mass significance") both in univariate and multivariate situations. There are some statistical security measures that will make such expeditions safer, but they should be used with care.

Use *more than one* set of severity scales. Examples of useful scales (preferred choices marked in boldface) are:

For schizophrenia
Brief Psychiatric Rating Scale (BPRS, mainly for the assessment of schizophrenia) (Bech et al., 1986; Overall, 1974; Overall and Gorman, 1962).
Comprehensive Psychopathological Rating Scale (CPRS) (Åsberg et al., 1973, 1978)
CPRS subscale for schizophrenia (Montgomery et al., 1987)
Schedule for the Assessment of Negative Symptoms (SANS) (Andreasen, 1989),
Positive And Negative Symptom Scale (PANSS) (Kay et al., 1989)
For depression
Hamilton Depression Rating Scale (Hamilton, 1960, 1967) which has been cross-culturally validated (Fava et al., 1982)
Hamilton Depression Rating Scale extracted from items in the Schedule of Affective Disorders and Schizophrenia (SADS) (Endicott et al., 1981).
Montgomery-Åsberg Depression Rating Scale (MADRS) (Montgomery and Åsberg, 1979)

For anxiety
Hamilton's Anxiety Rating Scale (Gjerris et al., 1983; Hamilton, 1959; Maier et al., 1988)

State or trait

If one attempts to demonstrate some PET characteristic to be a state variable or a marker of an episodic psychiatric disorder, then the variable should either change during the longitudinal course of the illness, *or* display a significant relation to a severity measure of some aspect of the illness. The finding of a significant correlation with more than one scale designed to measure the severity of the same "hidden structure" will strenghten the argument considerably. If no links with scales can be detected, the variable may candidate as a trait marker for some underlying vulnerability mechanism.

Elaboration

Elaborate on a tentative finding only if it *makes some sense* on a theoretical/ inductive level. The best validation is external and provided from a discipline in brain research which uses a different but related methodology.

Acknowledgements

This study was supported by the Swedish Medical Research Council (grants 6604 and 8461), the Fredrik and Ingrid Thuring Foundation, and the Söderström-König Foundation.

References

Andreasen NC (1989) The Scale for the Assessment of Negative Symptoms (SANS): conceptual and theoretical foundations. Br J Psychiatry 155: 49–58

Åsberg M, Kragh-Sørensen P, Mindham RH, Tuck JR (1973) International reliability and communicability of a rating scale for depression. Psychol Med 3: 458–465

Åsberg M, Perris C, Schalling D, Sedvall G (1978) The CPRS — development and applications of a psychiatric rating scale. Acta Psychiatr Scand 58 [Suppl]: 271

Bech P, Kastrup M, Rafaelsen OJ (1986) Mini-compendium of rating scales for states of anxiety, depression, mania, and schizophrenia, with corresponding DSM-III syndromes. Acta Psychiatr Scand 73: 7–37

Coccaro EF (1989) Central serotonin and impulsive aggression. Br J Psychiatry 155: 52–62

Endicott J, Cohen J, Fleiss J, Sarantakos S (1981) Hamilton depression rating scale: extracted from regular and change versions of the Schedule for Affective Disorders and Schizophrenia. Arch Gen Psychiatry 38: 98–103

Fava GA, Kellner R, Munari F, Pavan L (1982) The Hamilton depression rating scale in normals and depressives. Acta Psychiatr Scand 66: 26–32

Feighner J, Robins E, Guze S, Woodruff RA, Winokur G, Munoz R (1972) Diagnostic criteria for use in psychiatric research. Arch Gen Psychiatry 26: 57–63

Gjerris A, Bech P, Bøjholm S, Rafaelsen OJ (1983) The Hamilton Anxiety Scale. J Affect Disord 5: 163–170

Hamilton M (1959) The assessment of anxiety states by rating. Br J Med Psychol 32: 50–55

Hamilton M (1960) A rating scale for depression. J Neurol Neurosurg Psychiatry 23: 56–62

Hamilton M (1967) Development of a rating scale for primary depressive illness. Br J Soc Clin Psychol 6: 278–296

Kay SR, Opler LA, Lindenmayer J-P (1989) The Positive and Negative Syndrome Scale (PANSS): rationale and standardisation. Br J Psychiatry 155: 59–65

Klein DF (1978) A proposed definition of mental illness. In: Spitzer RL, Klein DF (eds) Critical issues in psychiatric diagnosis. Raven Press, New York, pp 41–71

Maier W, Buller R, Philipp M, Heuser I (1988) The Hamilton Anxiety Scale: reliability, validity, and senstivity to change in anxiety and depression. J Affect Disord 14: 61–68

Montgomery SA, Åsberg M (1979) A new depression scale designed to be sensitive to change. Br J Psychiatry 134: 322–389

Montgomery SA, Taylor P, Montgomery D (1987) Development of a schizophrenia scale sensitive to change. Neuropharmacology 17: 1061–1063

Overall JE (1974) The Brief Psychiatric Rating Scale in psychopharmacological research. In: Pichot P (ed) Modern problems of pharmacopsychiatry. Karger, Basel, pp 67–78

Overall JE, Gorman DR (1962) The Brief Psychiatric Rating Scale. Psychol Rep 10: 799–812

Roth M (1978) Psychiatric diagnosis in clinical and scientific settings. In: Akiskal HS, Webb WL (eds) Psychiatric diagnosis: exploration of biological predictors. SP Medical and Scientific Books, New York, pp 9–47

Spitzer RL, Endicott J, Robins E (1978) Research diagnostic criteria. Rationale and reliability. Arch Gen Psychiatry 35: 773–782

Author's address: Dr. H. Ågren, Department of Psychiatry, University Hospital, S-75185 Uppsala, Sweden

J Neural Transm (1992) [Suppl] 37: 27–38

Drug washout issues in studies of cerebral metabolism by positron emission tomography in psychiatric patients

J.-L. Martinot

Service Hospitalier Frédéric Joliot, CEA, Orsay, and Service de psychiatrie,
Hôpital A. Chenevier, Créteil, France

Summary. Many studies of brain glucose utilization by positron emission tomography attempt to describe the modifications of the brain activity during psychiatric diseases. A major difficulty in such studies is the necessity to assess patients free of pharmacological treatment, in order to relate the measured changes in glucose utilization to the pathopsychology, and not to a drug effect. In this paper are reviewed the arguments from the literature allowing to estimate the drug washout time for considering the patients as drug-free. The review is focussed on the known effects of the psychotrops on brain glucose utilization. This time is approximatively six months for the neuroleptics given orally, one month for antidepressants, and five and a half half-lives for benzodiazepines. Alternative research strategies for avoiding a long drug washout are mentioned, and ethical limitations are considered.

Introduction

During the last decade, many studies of the brain glucose metabolism have been performed with positron emission tomography (PET) in psychiatric patients in order to search for brain dysfunctions related to pathopsychology. However, patients were often studied during the course of a treatment with psychotropic drugs, or shortly after drug discontinuation. The necessity to focus on drug-naive or drug-free subjects when studying psychiatric patients with positron emission tomography is justified by the lack of information in regard to the effects of psychotropic drugs on cerebral metabolism, assessed by measuring brain glucose utilization. However, the recruitment of drug-naive or drug-free patients for PET studies of brain glucose metabolism is a difficult task because, in clinical practice, most patients are given neuroleptics right from the first psychotic symptoms. In an attempt to harmonize studies from different PET centers using the deoxyglucose method I will here review the arguments for the definition of drug-free patients, i.e. discuss the drug washout time that allow defining the patient as drug-free with respect to pharmacological effects on brain glucose utilization.

First, the issue of distinguishing the effects of psychotropic drugs from the effects of pathopsychology is considered, taking the measured regional cerebral metabolic rates for glucose ($rCMR_{Glc}$) as an index of cerebral activity. Second, the effects of the major psychotropic drugs used in psychiatry (i.e. neuroleptics, antidepressants, benzodiazepines) on the cerebral metabolism are reviewed, with emphasis on the regional cerebral metabolic rates for glucose. Drug washout intervals will be proposed based on the informations from this literature. These intervals, however, are to be considered with caution, since most of the information is either derived from animal studies, or from PET studies of psychiatric patients that were not designed to answer to the question of the time span necessary for the measured $rCMR_{Glc}$ to recover from a pharmacological perturbation. Third, the ethical limitations in studying drug-free patients wil be mentioned.

Distinguishing the effects of pathophysiology from those of drugs on brain metabolism

The effects of a therapeutic compound on $rCMR_{Glc}$ differ if it is administered acutely or chronically. This can possibly be explained by various time-dependent biochemical effects (e.g. receptor blockade > feedback activation > desensitization) related to the pharmacological action of the drug on the cerebral biochemical "plant". But this can also be linked to the changes in brain functional activity related to the change in behavior which goes along with the improvement of the psychiatric disease, thus questionning the specificity of the measures of brain glucose metabolism.

On the other hand, most psychiatric patients are usually treated for a long time, especially those suffering from a chronic illness. Excluding these patients from PET studies would render it difficult to obtain information on the influence of the chronicity on brain metabolism. Still, this kind of information is crucially needed because these patients are often the most impaired.

Longitudinal studies could constitute an alternative allowing to avoid a long drug washout. If one considers that the phenomenological particularities of the illness are reflected in the regional cerebral metabolic rates for glucose, one issue is to learn whether the observed $rCMR_{Glc}$ variations reflect (1) the changes in phenomenology induced by the therapeutic action of medication, rather than (2) the direct non-specific pharmacological effect of the drugs on brain metabolism, or (3) both of the above, representing, for instance, a selective effect of the drug on certain brain regions suggested to be functionnally abnormal in patients not taking medication. We foresee that this kind of problem induces a risk of circular reasoning, unless it is possible to study variations in only one of these terms.

Hence, a patient could be studied at different stages of his illness without modifications of the treatment. This is ethical for patients having a psychiatric condition which may change relatively fast. It could be the case

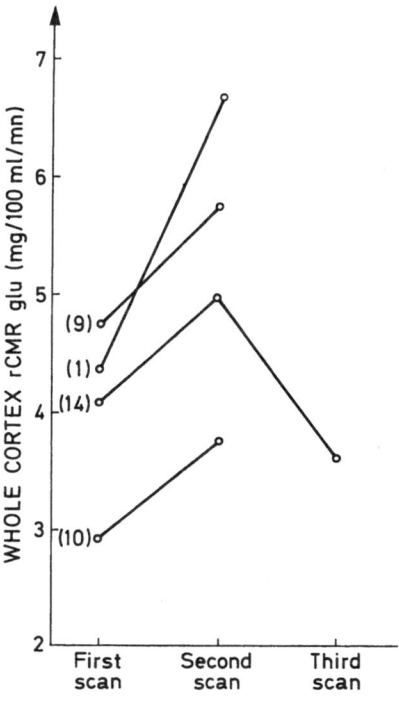

Fig. 1. Changes in global rCMR$_{Glc}$ (expressed in mg/100 ml/min) in four obsessive-compulsive (OC) patients. First scan: marked OC symptoms. Second scan: significant increase (paired t-test; p = 0.04) for the whole cortex glucose utilization, following clinical improvement. Third scan: OC symptoms relapse in patient 14

for a patient studied first while symptomatic, second after successful treatment, and third after a symptomatic relapse without modification of the previous treatment.

For instance, we have recently performed sequential measurements of the regional cerebral metabolic rates for glucose in four patients with obsessive-compulsive disorder (Martinot et al., 1990). The patients (Fig. 1) were subjected to a first PET study when the symptoms were are their peak, and to a second PET study several months later during clinical improvement; in addition, one patient was scanned for a third time during a relapse. Three patients were treated with antidepressants and benzodiazepines and one patient was only treated with behavioral psychotherapy. At the time of the first scan, the four patients had a global decrease in brain glucose utilization. At the second scan, however, all patients were in a recovering and had a global increase in glucose utilization, whichever treatment was used. Moreover, the patient scanned for a third time during a clinical relapse received exactly the same medications and doses as during the second scan, and again displayed a decreased rCMR$_{Glc}$. Performed on a larger scale, such studies could provide pathophysiological insights related to the changes in abnormal behaviors that occur as a manifestation of psychiatric disorders. Hence, longitudinal protocols are a way of avoiding the necessity of studying patients drug-free for a long time. Few studies of this kind are available.

Alternatively, it should be of interest to study sequentially the brain metabolism when a psychotropic compound is administered, that *fails* to change the phenomenology of the disease. If metabolic changes are observed, they could then be attributed to the medication. Besides, information is still scant on the particularities of non-responding patients as regards their cerebral metabolism. Also, for the anxiety disorders, little attention has been paid to the assessment of the effects of psychotherapeutic treatments (such as behavioral psychotherapy or relaxation techniques) on cerebral metabolism, which is a thrilling issue. For instance, studies of brain metabolism in obsessive-compulsive groups of patients, treated either by antidepressants or by behavioral psychotherapy, could shed light on the maybe differential impacts of these treatments on cerebral metabolism.

In conclusion: (1) PET protocols longitudinally assessing the effects on $rCMR_{Glc}$ of chronic treatments with benzodiazepines or antidepressants are critically needed, since the available literature on this topic is rare; (2) single cases studies in patients experiencing spontaneous variations of symptoms or benefitting from psychotherapy, and studies comparing non-responder chronic patients to responders, should be considered.

Time intervals for drug-free patients

One way to estimate the time intervals that would allow considering patients as drug-free is to review the effects of the common psychotropic drugs (neuroleptics, antidepressants, benzodiazepines) on the $rCMR_{Glc}$ as reported in studies on animals and humans. The effects of psychotropic drugs on certain other biological variables will also be mentioned.

Neuroleptics

Effects of neuroleptics on $rCMR_{Glc}$

A first group of studies were performed on laboratory animals using the autoradiographic method. Thus, such studies in rats using ^{14}C-deoxyglucose have been performed by McCulloch (1982) and Pizzolato (1984a,b, 1985, 1987). Pizzolato et al. reported that *acute* administration of haloperidol to rats significantly decreased glucose utilization by 60% in 59 brain regions examined, but produced large increases in the lateral habenula. However, *chronic* administration of haloperidol reduced glucose utilization to a lesser extent and in fewer regions than when used acutely. Interestingly, haloperidol had an effect on the metabolic activity of the central nervous system that was more widespread than would be predicted from the topography of the dopaminergic system. This may be due to the indirect propagation of the primary effects of haloperidol. Moreover, these authors showed (1987) that contrary to haloperidol, the benzamide sulpiride elevated glucose utilization in many brain regions, mainly related to the dopaminergic system. Finally, certain D_2 antagonists, but not D_1 antagonists, appear to modify the regulation of brain glucose metabolism in laboratory animals (Palacios and Wiederhold, 1985).

PET studies in humans have recently employed PET in schizophrenic patients, comparing their $rCMR_{Glc}$ to those of normal controls. Some investigators have studied drug-naive schizophrenics (Garnett et al., 1985; Volkow et al., 1986) with PET and reported no hypofrontality, at variance with the initial reports of Ingvar and Franzén (1974). The effect of neuroleptic treatments on $rCMR_{Glc}$ has also been assessed in schizophrenic patients according to a longitudinal design. Patients were usually studied twice, before and during treatment. A problem with some of these studies is that the investigated patients were chronic schizophrenics, and that the drug washout time was 1 month or less, rendering it difficult to assume that the measures at baseline were devoid of any residual effect of neuroleptic medication. Thus, Brodie et al. (1984) reported in 6 patients, drug-free since 15 days, a 25% increase of global CMR_{Glc} after treatment with thiothixene. However, the initial hypofrontality persisted despite clinical improvement. DeLisi et al. (1985) studied patients drug-free for at least 15 days, treated with chlorpromazine, fluphenazine, haloperidol, clozapine, or thioridazine, the duration of treatment being 3 to 32 months, until clinical stabilization. They reported an increase of cortical glucose utilization and a relative increase in the caudate nuclei. Furthermore, the antero-posterior cortical gradients did not change. On the whole, these two studies indicated a global increase of the cerebral glucose utilization, especially marked in the caudate nuclei, after a period of successful use of neuroleptics. However, the washout time was short.

Wolkin et al. (1985) confirmed the global increase related to chronic neuroleptic treatment in schizophrenics, and stated that this elevation was more marked in the posterior cortical areas. However, in this study as well, the metabolic gradient from front to back appeared to remain unaffected by neuroleptic treatment in patients having been studied both off and on chronic treatment. Drug washout was longer than or equal to 17 days. In two other studies, no changes in brain glucose metabolism following acute (Volkow et al., 1986) or chronic (Gur et al., 1987) neuroleptic administration were found. No difference in $rCMR_{Glc}$ between drug-naive and drug washout conditions was reported in these patients.

Wik et al. (1989) studied specifically the effect of sulpiride or chlorpromazine on $rCMR_{Glc}$ in schizophrenics. They controlled for drug compliance, fixed doses, and used a fixed length of treatment. Among 17 patients, 3 were off neuroleptics since less than 6 months; the others were off these kinds of drugs for a longer period of time. A significant change in glucose metabolism related to drug treatment was only found in one brain region — an increased metabolic rate in the right lentiform nucleus. The authors remarked that effects of neuroleptics on rCMR$_{Glc}$ largely seem to become less important when studies are better controlled.

Time-related effects of neuroleptic withdrawal on other biological parameters

Although the correlation between the following parameters and the cerebral glucose consumption is unknown, they provide indications as to the duration of the pharmacological effects of neuroleptics after discontinuation.

D_2 *dopamine receptors.* The in vivo state of D_2 receptors after neuroleptic withdrawal have recently been assessed. A study by Baron (1989) using ^{76}Br-bromospiperone as ligand stated that following neuroleptic withdrawal there is a sharp increase in the available binding site percentage. This increase would indicate a return to a 100% availability of the D_2 striatal dopamine receptors within six to twelve days. Moreover, increase of the D_2 receptor number due to chronic neuroleptization is well documented in animal studies, but the duration of this "up-regulation" is unknown in man. However, in a recent post-mortem study of schizophrenic patients, Kornhuber (1989) showed that the density of D_2 receptors was normal if the neuroleptic treatment had been discontinued 3 months before death, suggesting that the alleged up-regulation normalizes within this time.

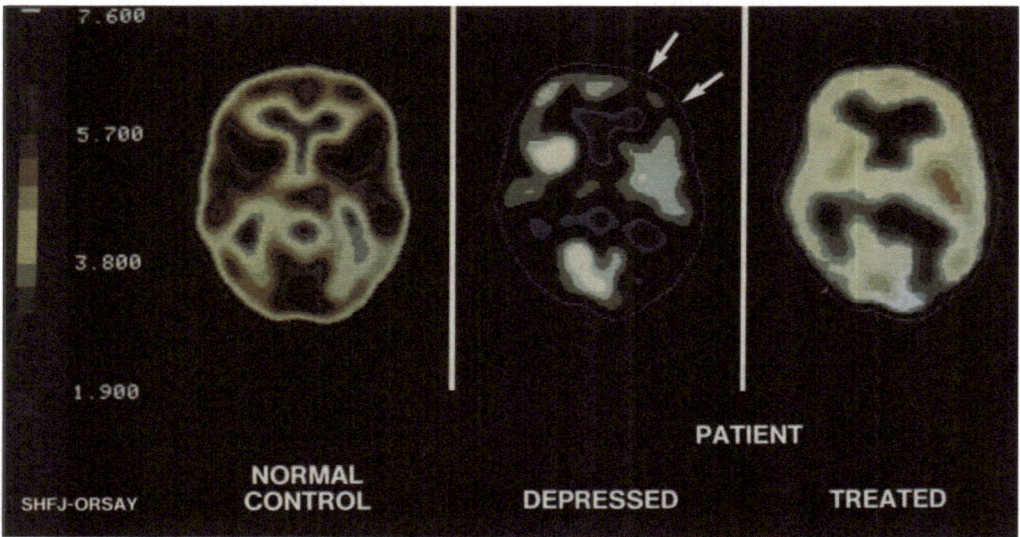

Fig. 2. Cerebral slices showing rCMR$_{Glc}$ 40 millimeters above the orbito-meatal line in a control subject (left) and in a patient studied in the depressed state (center) and after successful treatment (right). Note the global decrease of glucose utilization in the depressed state, more marked in the left prefrontal region (arrows). After treatment, the frontal asymmetry diseapears, and the glucose metabolic rate in the whole cortex increases but non significantly

Neuroleptic plasma concentration. Three studies report a longer persistance of neuroleptics in the plasma than is usually considered. Hubbard (1987) reported that the slow disappearance of neuroleptic effects after discontinuation of treatment could be related to an ultra-low level phase of elimination of the drug from plasma. The half-life of haloperidol was estimated to be 13 hours on the first day of withdrawal, but was estimated to be 21 days eleven days after withdrawal. For the depot neuroleptics, 6 months after withdrawal, plasma levels of fluphenazine decanoate were still detectable (Wistedt et al., 1982).

Interestingly, a single dose of haloperidol can provoke long-lasting physiological effects. Campbell et al. (1985) showed that its antidopaminergic effects can last 20–40 days in laboratory animals. This effect was evaluated in rats by the decrease of their stereotyped behavioral responses to an acute dose of apomorphine.

In summary, in the absence of PET studies designed to measure the duration of the effects of neuroleptics on cerebral glucose regional consumption, the available literature suggests that patients can be considered neuroleptic-free if they have been off neuroleptics for 6 months. It would appear that patients treated with depot neuroleptics can only be considered neuroleptic free when they have been off neuroleptics for at least one year, but there is a lack of data on this issue.

Antidepressants

Effects of antidepressants on brain glucose metabolism

In *animals*, the acute administration of clomipramine did not modify the in vivo global $rCMR_{Glc}$ in anesthetized dogs, although $rCMR_{O_2}$ decreased significantly (Sari et al., 1975).

In *humans*, few PET studies have assessed the effects of antidepressants on $rCMR_{Glc}$ in psychiatric samples (Baxter et al., 1985, 1989; Martinot et al., 1990a; Benkelfat et al., 1990). Increases of $rCMR_{Glc}$ in various brain regions and particularly in the left frontal region have been pointed out in patients treated for depressive episodes. These increases are probably linked to the changes in affective states due to treatment. Thus, we investigated (1990a) the resting-state metabolic rates for glucose in ten severely depressed patients before and after treatment with tricyclic antidepressants, and compared the rates to those of ten control subjects of similar age. A significant left-right prefrontal asymmetry was present in the patients before, but not after successful treatment, suggesting that medication can reduce this asymmetry (Fig. 2). However, a significant hypofrontality and a whole cortex hypometabolism were found in the patients in the depressed state, persisting in the treated state despite clinical improvement which would suggest that these abnormalities were state-independent. Still, these studies were not designed to discriminate between the effects of clinical recovery and the effects of medications. Besides, in the anxiety disorder spectrum, Baxter et al. (1986) indicated a relative increase of caudate metabolism in patients with obsessive-compulsive disorder treated with trazodone. Also in obsessive-compulsive disorder, Benkelfat et al. (1990) studied eight patients before and after treatment with the tricyclic antidepressant clomipramine. Comparisons of local $rCMR_{Glc}$ for both groups showed a relative decrease in regions of the orbital cortex and of the left caudate, and an increase in other areas of the basal ganglia, including the right anterior putamen. Only differences in the left caudate differentiated the patients who responded well to clomipramine from those who were either poor or partial responders.

Effects of antidepressant withdrawal on other biological parameters

Only after a period of 4 weeks of washout from treatment with imipramine the B_{max} of [^{3}H]imipramine returns to baseline levels. As a contrast, 1 week of treatment with maprotiline or with amineptine did not change [^{3}H]5-HT uptake or [^{3}H]imipramine binding (Poirier, 1987). Uptake measures of serotonin and noradrenalin were normalized within 1 to 2 weeks after zimeldin withdrawal, whereas the effects of clomipramine persisted for 3 to 4 weeks (Ross, 1983). Also, rebound effects should be considered after abrupt discontinuation of tricyclic antidepressant drugs — plasma levels of the noradrenalin metabolite 3-methoxy-4-hydroxyphenylethylamine were increased to a peak 15 to 20 days following discontinuation (Charney, 1982). Rebound activation of post-dexamethazone plasma cortisol level normalizes within 21 days after withdrawal of antidepressants (Kraus, 1987). Finally, one must keep in mind that intracellularly, psychotropic drugs can modify the transcription of mRNA for more than 18 days in animals (Faucon Biguet, 1986).

On the whole, a washout period of 4 weeks may be needed for patients treated with clomipramine. Various other antidepressants have biological effects that persist more than 15 days after withdrawal. Consequently, a patient would be considered antidepressant-free if the drug washout period is equal to or longer than 4 weeks.

Benzodiazepines

In autoradiographic studies in animals, acute administration of diazepam reduced the $rCMR_{Glc}$ by up to 30% in several subcortical structures, but no changes were observed at the cortical level. Chronic treatment with diazepam produced slight reductions in areas such as lateral thalamic nuclei (Ableitner, 1987). Other gabaergic substances such as meprobamate and phenobarbital decreased $rCMR_{Glc}$ in a dose-dependent manner (Ableitner, 1987).

Table 1. Time of complete washout for various benzodiazepines (from Boulenger, 1986)

International drug name	Complete drug washout (days)
Alprazolam	4
Bromazepam	4
Chlordiazepoxide	13
Clobazam	8
Clorazepate	17
Clotiazepam	2.5
Desmethyldiazepam	17
Diazepam	17
Estazolam	5
Flunitrazepam	7
Lorazepam	5
Medazepam	17
Nitrazepam	8
Oxazepam	4
Prazepam	17
Temazepam	2.5
Triazolam	1

Few published PET studies have assessed the effects of benzodiazepines on human $rCMR_{Glc}$. According to Foster (1987), *acute* administration of *sedative* doses of diazepam to Alzheimer patients would depress overall glucose utilization by 20%. Only one study assessed the effects of *chronic* treatment with benzodiazepines on regional glucose metabolism: decreases in glucose metabolic rates were found only in visual cortex, but relative increases were reported in the basal ganglia and thalamic areas (Buchsbaum et al., 1987).

As regards the design of the protocols of PET studies, it should be noticed that it is unethical to withhold treatment from a patient suffering from pathological anxiety. Low doses of benzodiazepines with a short half-life should be used when clinical conditions render it necessary.

Otherwise, some authors state that patients treated with benzodiazepines can be considered drug-free after a washout period of 5.5 half-lives of the compound (Boulenger, 1986). As the half-life of benzodiazepines varies depending of the compound, estimations of the time period needed to achieve a complete drug washout are indicated in Table 1. Of course, benzodiazepines must be slowly tapered in order to avoid withdrawal symptoms.

Ethical considerations

The following considerations aim at pointing out some issues which may arise when designing the protocol of a study using PET. They are not to be taken as generally accepted rules. In all cases, the protocol of a PET research study must be submitted to the approval of an ethics committee.

One first point is that these studies provide no direct individual benefits in psychiatric patients because they have neither diagnostic and prognostic value today, nor will they guide drug treatment. Nevertheless, they yield information either on functional anatomy of mental diseases or on the drug-induced changes in cerebral metabolism.

In studying "drug-clean" patients one may be confronted to the issue of whether it is feasible to perform a drug washout or to potentially delay an effective treatment. Despite the need to study *drug-free* patients, i.e. patients having a sufficient period of drug washout, it is of course unethical to interrupt a successful drug treatment. However, interrupting a treatment may be indicated after it has failed to improve symptoms, or if there is a medical reason to reevaluate the treatment (uncontrollable side effects, for instance). Also, one may discuss the need for prolonged treatment if the symptoms have been controlled for a lengthy period of time. This raises the question of studying pathological traits, rather than states.

PET studies of *drug-naive* patients are needed in order to describe the changes in cerebral metabolism related to the illness. However, delaying the treatment is unethical unless a period of time is needed to assess the diagnosis. For instance, several days may be needed to determine the clinical type of certain delusions. Also, patients may have a spontaneous decrease in their symptoms due to the beneficial effect of hospitalization, and several days may be needed to reevaluate the symptomatology and even the diagnosis after this initial improvement.

Hence, acceptable delays may be useful in drug-naive patients, in order to ensure a valid diagnosis and to give the most suitable drug treatment.

References

Ableitner A, Herz A (1987) Influence of meprobamate and phenobarbital upon local cerebral glucose utilization: parallelism with effects of the anxiolytic diazepam. Brain Res 403: 82–88

Ableitner A, Wüster M, Herz A (1985) Specific changes in local cerebral glucose utilization in the rat brain induced by acute and chronic diazepam. Brain Res 359: 49–56

Baxter LR, Phelps ME, Mazziotta JC, Schwartz JM, Gerner RH, Selin CE, Sumida RM (1985) Cerebral metabolic rates for glucose in mood disorders. Arch Gen Psychiatry 42: 441–447

Baxter LR, Phelps ME, Mazziotta JC, Guze BH, Schwartz JM, Selin CE (1987) Local cerebral metabolic rates in nondepressed patients with obsessive-compulsive disorder: a comparison with rates in unipolar depression and normal controls. Arch Gen Psychiatry 44: 211–218

Baxter LR, Schartz JM, Phelps ME, Mazziotta JC, Guze BH, Selin CE, Gerner RH, Sumida RM (1989) Reduction of prefrontal cortex glucose metabolism common to three types of depression. Arch Gen Psychiatry 46: 243–250

Benkelfat C, Nordahl TE, Semple W, King C, Murphy DL, Cohen RM (1990) Local cerebral glucose metabolic rates in obsessive compulsive disorder. Patients treated with clomipramine. Arch Gen Psychiatry 47: 840–848

Boulenger JP (1986) Caractéristiques pharmacocinétiques des benzodiazépines. Act Méd Inter-Psychiatrie 36: 913–914

Brodie JD, Christman DR, Corona JF, Fowler JS, Volkow ND, Wolf AP, Wolkin A (1984) Patterns of metabolic activity in the treatment of schizophrenia. Ann Neurol 15: 166–169

Buchsbaum MS, Wu J, Haier R, Hazlett E, Ball R, Katz M, Sokolski K, Lagunas-Solar M, Langer D (1987) Positron emission tomography assessment of effects of benzodiazepines on regional glucose metabolic rate in patients with anxiety disorder. Life Sci 40: 2393–2400

Campbell A, Baldessarini RJ, Teicher MH, Kola NS (1985) Prolonged antidopaminergic action of single doses of butyrophenones in the rat. Psychopharmacology 87: 161–166

Charney DS, Heninger GR, Sternberg DE, Landis H (1982) Abrupt discontinuation of tricyclic antidepressant drugs: evidence for noradrenergic hyperactivity. Br J Psychiatry 141: 377–386

DeLisi LE, Holcomb HH, Cohen RM, Pickar D, Carpenter W, Morihisa JM, King AC, Kessler R, Buchsbaum MS (1985) Positron emission tomography in schizophrenic patients with and without neuroleptic medication. J Cereb Blood Flow Metab 5: 201–206

Faucon Biguet N, Buda M, Lamouroux A, Samolyk D, Mallet J (1986) Time course of the changes of TH mRNA in rat brain and adrenal medulla after a single injection of reserpine. EMBO J 5: 287–291

Foster N, van der Speck AFL, Aldrich MS, Berent S, Hichwa RH, Sackellares JC, Gilman S, Agranoff BW (1987) The effect of diazepam sedation on cerebral glucose metabolism in Alzheimer's disease as measured using positron emission tomography. J Cereb Blood Flow Metab 7: 415–420

Garnett ES, Nahmias C, Cleghorn G (1985) Pattern of local cerebral glucose metabolism in untreated schizophrenics. J Cereb Blood Flow Metab 5 [Suppl] 1: 179–180

Gur RE, Resnick SM, Gur RC, Alavi A, Caroff S, Kushner M, Reivich M (1987) Regional brain function in schizophrenia. Arch Gen Psychiatry 44: 126–129

Hubbard JW, Ganes D, Midha KK (1987) Prolonged pharmacological activity of neuroleptic drugs. Arch Gen Psychiatry 44: 99–100

Ingvar DH, Franzén G (1974) Abnormalities of cerebral blood flow distribution in patients with chronic schizophrenia. Acta Psychiatr Scand 50: 425–462

Kornhuber J, Riederer P, Reynolds GP, Beckmann H, Jellinger K, Gabriel E (1989) ^{3}H-spiperone binding in post mortem brains from schizophrenic patients: relationship to neuroleptic drug treatment, abnormal movements, and positive symptoms. J Neural Transm 75: 1–10

Kraus RP, Hux M, Grof P (1987) Psychotropic drug withdrawal and the dexamethasone suppression test. Am J Psychiatry 144: 82–85

McCulloch J, Savaki HE, Sokoloff L (1982) Distribution of effects of haloperidol on energy metabolism in the rat brain. Brain Res 243: 81–90

Martinot JL, Allilaire JF, Mazoyer BM, Hantouche E, Huret JD, Legaut-Demare F, Deslauriers AG, Hardy P, Pappata S, Baron JC, Syrota A (1990) Obsessive-compulsive disorder: a clinical, neuropsychological and positron emission tomography study. Acta Psychiatr Scand 82: 233–242

Martinot JL, Hardy P, Feline A, Huret JD, Mazoyer B, Attar-Levy D, Pappata S, Syrota A (1990a) Left prefrontal glucose hypometabolism in the depressed state: a confirmation. Am J Psychiatry 147: 1313–1317

Palacios JM, Wiederhold KH (1985) Dopamine D_2 receptor agents, but not dopamine D_1, modify brain glucose metabolism. Brain Res 327: 390–394

Pizzolato G, Soncrant TT, Rapoport S (1984) Haloperidol and cerebral metabolism in the conscious rat: relation to pharmacokinetics. J Neurochem 43: 724–732

Pizzolato G, Soncrant TT, Larson DM, Rapoport SI (1985) Reduced metabolic response of the rat brain to haloperidol after chronic treatment. Brain Res 335: 1–9

Pizzolato G, Soncrant TT, Larson DM, Rapoport SI (1987) Stimulatory effect of the D_2 antagonist sulpiride on glucose utilization in dopaminergic regions of rat brain. J Neurochem 49: 631–638

Poirier MF, Galzin AH, Lôo H, Pimoule C, Segonzac A, Benkelfat C, Sechter D, Zarifian E, Schoemaker H, Langer SZ (1987) Changes in [^{3}H]5-HT uptake and [^{3}H]imipramine binding in platelets after chlorimipramine in healthy volunteers. Comparison with maprotiline and amineptine. Biol Psychiatry 22: 287–302

Ross SB, Åberg-Wistedt A (1983) Inhibitors of serotonin and noradrenalin uptake in human plasma after withdrawal of zimelidine and clomipramine treatment. Psychopharmacology 79: 298–303

Sari A, Fukuda Y, Sakabe T, Maekawa T, Toshizo I (1975) Effects of psychotropic drugs on canine cerebral metabolism and circulation related to EEG. Diazepam, clomipramine, and chlorpromazine. J Neurol Neurosurg Psychiatry 38: 838–844

Volkow ND, Brodie JD, Wolf A, Angrist B, Russel J, Cancro R (1986) Brain metabolism in patients with schizophrenia before and after acute neuroleptic administration. J Neurol Neurosurg Psychiatry 49: 1199–1202

Wik G, Wiesel FA, Sjögren I, Blomqvist G, Greitz T, Stone-Elander S (1989) Effects of sulpiride and chlorpromazine on regional cerebral glucose metabolism in schizophrenic patients as determined by positron emission tomography. Psychopharmacology 97: 309–318

Wistedt B, Jørgensen A, Wiles D (1982) A depot neuroleptic withdrawal study. Plasma concentration of fluphenazine and flupenthixol and relapse frequency. Psychopharmacology 78: 301–304

Wolkin A, Jaeger J, Brodie JD, Wolf AP, Fowler J, Rotrosen J, Gomez-Mont F, Cancro R (1985) Persistence of cerebral metabolic abnormalities in chronic schizophrenia as determined by positron emission tomography. Am J Psychiatry 142: 564–571

Author's address: Dr. J.-L. Martinot, Psychiatric Department, Albert Chenevier Hospital, 70 rue de Mesly, F-94010 Créteil, France

J Neural Transm (1992) [Suppl] 37: 39–52

The FDG model and its application in clinical PET studies

K. Wienhard

Max-Planck-Institut für Neurologische Forschung, Köln (Lindenthal),
Federal Republic of Germany

Summary. The FDG method, as it is applied in clinical PET studies is reviewed. The influence of different implementations of the method and instrumental inaccuracies on the values of cerebral metabolic rate of glucose is discussed. For the comparison of the results between different groups standardized procedures are recommended.

Introduction

Positron emission tomography (PET) of $[^{18}F]$2-fluoro-2-deoxy-D-glucose (FDG) is a widely used technique for local estimation of glucose metabolism in vivo. The FDG-model is an adaption of the $[^{14}C]$2-deoxy-D-glucose-model originally developed by Sokoloff et al. (1977) for quantitative auto-radiography in animals, to measure glucose metabolism, mainly in brain, with PET in man (Reivich et al., 1979). Presently, the FDG-method is used extensively in many PET-centers for clinical studies. However, the implementation of this method: the protocols, the instruments used to measure brain and blood activity concentration, the analysis of the data and the modelling approaches, may differ considerably among laboratories. Therefore, the comparison of absolute metabolic rates between different groups and laboratories makes a thorough discussion of these topics necessary. If possible, recommendations for standardized procedures should be made and agreed on in the future.

Theory

Based on the biochemical behavior of deoxyglucose in brain Sokoloff et al. (1977) derived a model as illustrated in Fig. 1. Deoxyglucose (DG) and glucose in the plasma share and compete for a common carrier in the blood-brain barrier (BBB) for transport from plasma to brain. In brain tissue they compete either for the carrier for transport back from brain to plasma or for the enzyme hexokinase, which phosphorylates them irreversibly to

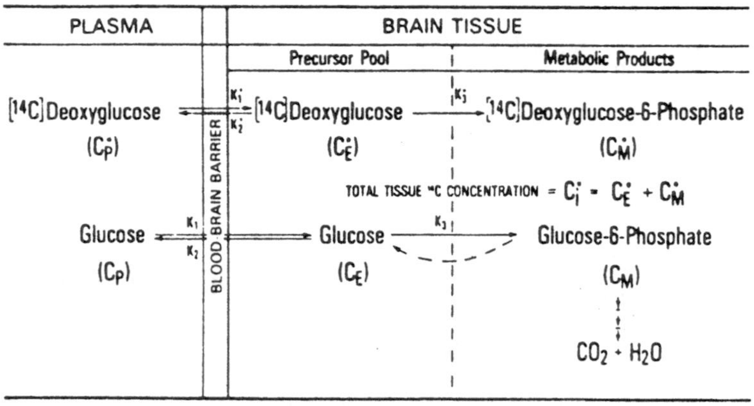

$$\text{CMRGlc} = \frac{C_P\left[C_I^*(T) - k_1^* e^{-(k_2^* + k_3^*)T} \int_0^T C_P^*(t) e^{(k_2^* + k_3^*)t} dt\right]}{LC\left[\int_0^T C_P^*(t)dt - e^{-(k_2^* + k_3^*)T} \int_0^T C_P^*(t) e^{(k_2^* + k_3^*)t} dt\right]}$$

Fig. 1. Compartmental model to describe behavior of glucose and FDG in brain tissue

deoxyglucose-6-phosphate (DG-6-P) and glucose-6-phosphate (G-6-P). G-6-P is then metabolized further eventually to carbon dioxide and water. On the other hand, DG-6-P is no substrate for any enzyme known to be present in brain tissue and is therefore trapped and accumulates as it is formed. The possibility of hydrolysis back by glucose-6-phosphatase activity can be incorporated into the model but the activity of this enzyme is believed to be very low in mammalian brain.

The DG model depends on several assumptions:

1. The tissue compartment should be homogeneous with respect to blood flow, transport rates and concentrations.
2. Glucose metabolism is in a steady state. The rate of glucose utilization, the plasma glucose concentration and the concentrations of all the substrates and intermediates of the glycolytic pathway are constant during the measurement.
3. DG and DG-6-P are present in tracer amounts.
4. Tissue extraction fraction of glucose and DG from plasma is small.

From the differential equations describing the compartment model in Fig. 1 the concentrations C_E and C_M of radioactivity can be expressed as

$$C_E(t) = K_1 e^{-(k_2 + k_3)t} \otimes C_p^*(t) \qquad C_M(t) = \frac{K_1 k_3}{k_2 + k_3}\left[1 - e^{-(k_2 + k_3)t}\right] \otimes C_p^*(t)$$

$$(1)$$

and total activity in tissue $C_i(t)$ is given by

$$C_i(t) = C_E(t) + C_M(t)$$

$$= K_1 \left[\frac{k_3}{k_2 + k_3} \int_o^t C_p^* (t') \, \mathrm{dt'} + \frac{k_2}{k_2 + k_3} exp \left[-(k_2 + k_3)t \right] \right.$$

$$\left. \int_o^t C_p^* (t') \, exp \left[(k_2 + k_3)t' \right] dt' \right] \qquad (2)$$

where $C_i(t)$ is the total amount of tracer in tissue, $C_p^*(t)$ is the capillary plasma FDG concentration as a function of time, K_1, k_2 and k_3 are first order rate constants and $\otimes$ denotes the operation of convolution.

The metabolic rate is given by

$$CMR_{Glc} = \frac{C_p}{LC} \frac{K_1 k_3}{k_2 + k_3} \qquad (3)$$

with C_p the concentration of glucose in plasma. LC represents the "lumped constant" accounting for the differences in transport and phosphorylation between glucose and deoxyglucose. Originally it was assumed that DG-6-P is essentially trapped in tissue for the duration of the period of measurement, because the activity of phosphatase, which hydrolyzes DG-6-P is low in brain tissue. Phelps and coworkers (1979) took into account the possible dephosphorylation reaction by extending Sokoloff's model to incorporate a fourth rate constant, representing the slow process of dephosphorylation. There are several methods to apply the Sokoloff model to derive local cerebral metabolic rates of glucose (CMR_{Glc}) from measurements of radioactivities in brain with PET.

Single scan methods

The most widely used method of measuring CMR_{Glc} is the so-called in vivo autoradiographic approach. It is based on the original Sokoloff operational equation:

$$CMR_{Glc} = \frac{C_p}{LC} \frac{K_1 k_3}{k_2 + k_3} \approx \frac{C_p}{LC} \frac{\bar{K_1} \bar{k_3}}{\bar{k_2} + \bar{k_3}} \left[\frac{C_i(T) - \bar{C}_E(T)}{\bar{C}_M(T)} \right] \qquad (4)$$

$\bar{K_1}$, $\bar{k_2}$ and $\bar{k_3}$ are mean values of the rate constants in a population of control subjects; $\bar{C}_E(T)$ and $\bar{C}_M(T)$ the radioactivity concentrations as calculated from the above formulas with $\bar{K_1}$ and $\bar{k_2}$ and $\bar{k_3}$ as set parameters. $C_i(T)$ is the PET-measured total tissue radioactive concentration at time T after injection. The extension of the model to include dephosphorylation as proposed by Phelps et al. (1979) modifies the computation of $\bar{C}_E(T)$ and $\bar{C}_M(T)$ by adding an average $\bar{k_4}$ in the model equation (Phelps et al., 1979). Eq. 4 expresses CMR$_{Glc}$ as a linear function of the measured PET-data $C_i(T)$. The assumption is, that $C_M(t)$ is proportional to the metabolic rate while $C_E(t)$ is independent of it. However, the formula gives only an approximation of the true metabolic rate because of the deviations of the individual subjects rate constants from the population mean rate constants. Because it has been shown that the formula is quite sensitive to these deviations (Huang et al., 1980; Hawkins et al., 1981) and may in pathological cases even lead to erroneous negative values for CMR_{Glc} (Heiss et al., 1983), alternative formulas based on different hypotheses have been developed by several authors. Hutchins et al. (1984) proposed to assume $C_i(T)$ to be proportional to CMR_{Glc}, which leads to the formula

$$CMR_{Glc} \approx \frac{C_p}{LC} \frac{\bar{K}_1 \bar{k}_3}{\bar{k}_2 + \bar{k}_3} \left[\frac{C_i(T)}{\bar{C}_E(T) + \bar{C}_M(T)} \right] \qquad (5)$$

Because the quantities $C_E(T)$ and $C_M(T)$ are linearly dependent on K_1, CMR_{Glc} estimations using this formula are independent of the K_1 value. The Hutchins approach is therefore identical to the proposed K_1 normalization procedure of Heiss et al. (1983). Brooks (1982) observed that the calculation of the total model predicted tissue radioactivity concentration $C_i(T) = C_E(T) + C_M(T)$ could be rewritten as the sum of two functions, a slowly decaying one $C_s(T)$ and a rapidly decaying one $C_R(T)$, both being the product of a constant coefficient and the convolution of the plasma time-activity function with an exponential. The constant coefficient of $C_s(T)$ being nearly equal to the ratio $K_1 k_3 / (k_2 + k_3)$ suggested that $C_s(T)$ would be proportional to CMR_{Glc}, and would make the following formula less sensitive to deviations of the rate constants from the normal mean values:

$$CMR_{Glc} \approx \frac{C_p}{LC} \frac{\bar{K}_1 \bar{k}_3}{\bar{k}_2 + \bar{k}_3} \left[\frac{C_i(T) - \bar{C}_R(T)}{\bar{C}_s(T)} \right] \qquad (6)$$

A very simplified equation was proposed by Rhodes et al. (1983)

$$CMR_{Glc} \approx \frac{C_p}{LC} \frac{\bar{k}_1 \cdot \bar{k}_3}{\bar{k}_2 + \bar{k}_3} \cdot \frac{C_i(T)}{\displaystyle\int_o^T C_p^*(t')\, dt'} \qquad (7)$$

To take into account the individual variations of rate constants in certain pathological conditions a normalization method was proposed by Wienhard et al. (1985) which employs the measured tissue tracer concentration itself as a corrective. Depending on what is known about the usual variation of the rate constants in certain pathologies the discrepancy between PET-measured tissue activity and calculated tissue activity $C_i(T)$ using standard rate constants in Eq. 2 can be used to normalize the rate constants in such a way that when inserted in Eq. 2 the measured tissue activity is reproduced.

Dynamic method

The dynamic approach has been implemented in two different ways. The first depends on a graphical analysis that is linear if $k_4 = 0$ and that provides the $K_1k_3/(k_2 + k_3)$ ratio by linear fitting. As demonstrated by Patlak et al. (1983) under quite general assumptions and by Gjedde et al. (1985) for FDG, a graph of the ratio of the tracer tissue concentration C_i to the plasma concentration C_p^* versus the ratio of the arterial plasma concentration time integral to C_p^* yields a curve that eventually approaches a straight line with the slope $K_1k_3/(k_2 + k_3)$. This relation is easily derived from Eq. 2. Under steady state conditions, i.e., at times t when plasma concentration C_p^* can be considered constant compared to the strong time dependence of the exponential factor $\exp[-(k_2 + k_3)t]$, Eq. 2 can be written as

$$\frac{K_1k_3}{k_2 + k_3} = \frac{C_i(t)/C_p^*(t) - K_1k_2/(k_3 + k_3)^2}{\int_o^t C_p^*(t')\,dt'/C_p^*(t)} \tag{8}$$

The graphical plot is never quite linear (because $k_4 > 0$ or influences of tissue heterogeneity) so that the estimation of CMR_{Glc} by this method remains uncertain to some degree.

Dynamic curve fitting after administration of labeled DG probably remains the most accurate way of determining CMR_{Glc}. The individual rate constants of the three compartment model can be determined as well by utilizing the potential of PET to take repeated measurements in rapid succession of local radioactivity in the same tissue, thus characterizing the complete net uptake of ^{18}F in the brain, starting at tracer injection (Fig. 2). A least-squares fit to those time-activity data then yields the parameters of the model equation. With a multiring PET system allowing simultaneous recordings of the tracer tissue concentration in such a number of slices as to cover the whole brain, this procedure is readily feasible on a routine basis. For optimum reproducibility of results, however, besides adequate control of experimental conditions during the examination, a well-devised fit stratagem is essential.

The metabolic factor, i.e., the term $K_1k_3/(k_2 + k_3)$ is determined by the later portion ($t > 15\,\text{min}$) of the time activity curve, which is equally well fitted by an infinite number of parameter sets. However, as can be seen from Eq. 2, K_1 is uniquely determined by the very early part of the uptake curve as

$$\lim_{t \to 0} \frac{C_i(t)}{\int_o^t C_p^*(t')\,dt'} = K_1 \tag{9}$$

Hence, the initial slope of the uptake curve defines K_1, thereby reducing the number of parameters that actually need to be estimated by fitting Eq. 2 to the time activity data by one and, in principle, permitting a more precise fit solution.

However, for this approach correction for radioactivity within the vascular space is critical in the early phases for accurate curve fitting. This can be performed either by means of an additional scan with labeled carbon

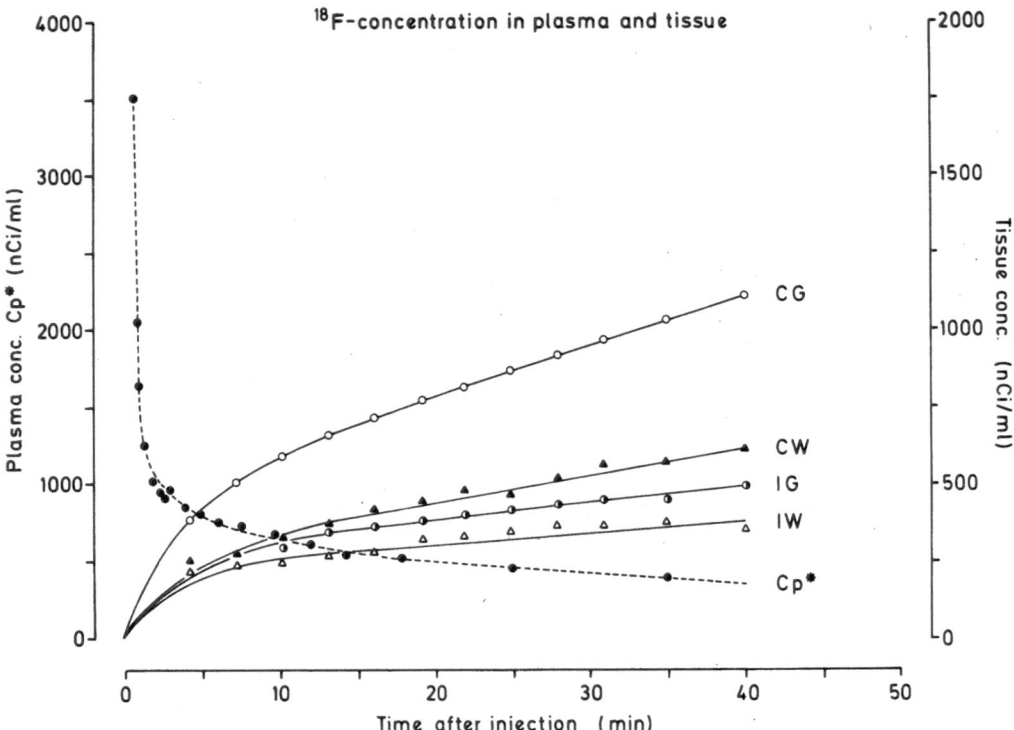

Fig. 2. Characteristic time courses of decay corrected isotope concentrations in plasma C_p^* (left scale) and in various brain regions (right scale). Respective curves were least-squares fitted according to Eq. 2

monoxide, or by including the vascular space in the model and taking it into account in the fitting procedure (Hawkins et al., 1986). In this case it is necessary to sample brain radioactivity rapidly with PET and the input function must be well characterized. Fast or (better) continuous sampling of blood is necessary to define peak activity accurately. The time shift between brain and the blood sampling site is critical and can be either estimated by detection from the brain peak by actual monitoring of the count rates or by fitting it in the curve-fitting procedure (Mazoyer et al., 1986).

There are several reports on pixel-by-pixel estimations of the individual deoxyglucose rate constants (Baron et al., 1984; Sasaki et al., 1986; Herholz, 1988) but these procedures are very time consuming and did not take into account all of the above factors.

There is only limited indication that knowledge of individual DG rate constants in the brain on a regional basis is of significant physiological interest (Wienhard et al., 1985). In cases where the use of normal rate constants in the autoradiographic operational equation may lead to considerable errors in CMR_{Glc} (especially in pathologies with largely reduced metabolism) or in hyperglycemia curve fitting is required. It should be realized that the ratio $K_1 k_3/(k_2 + k_3)$ is much more accurately estimated

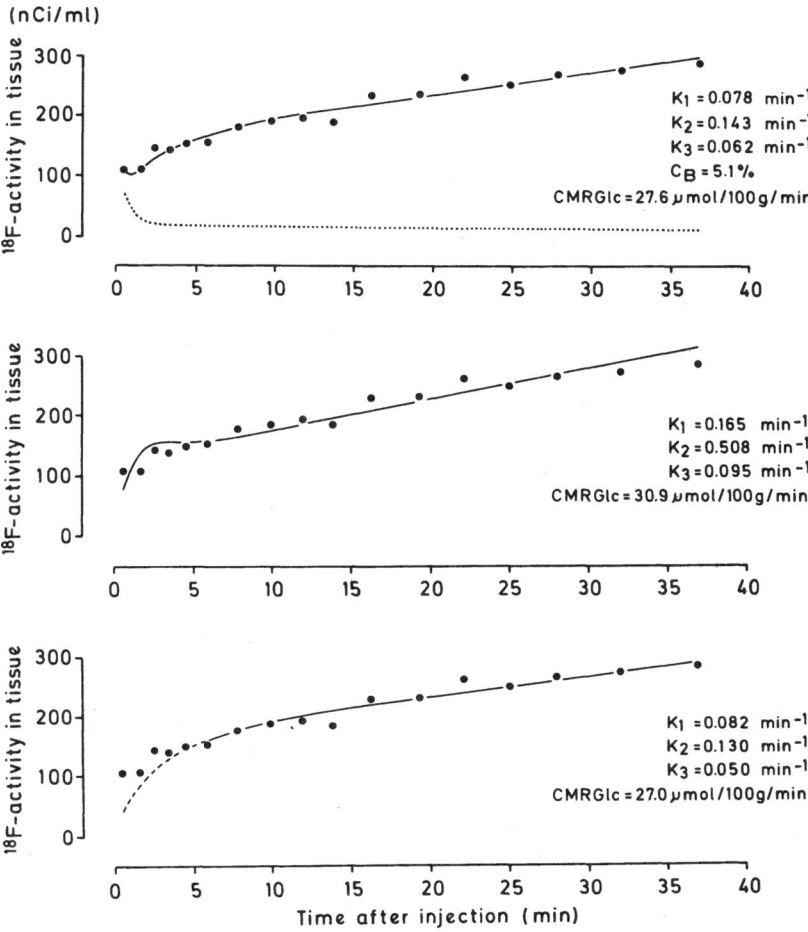

Fig. 3. Different least-squares fits for the same data set. The upper and middle graphs show fit results obtained when all data points are used in the procedure; for the lower graph, start of fit was set at t = 6 min after injection. Only in the curve fitting represented by the upper graph was fractional blood volume C_B (dotted curve) included as an additional parameter

than the individual rate constants. It was proposed (Wienhard et al., 1985) that in this case the first 3 min of the brain curve may be discarded in order to limit the influence of the vascular fraction correction and attenuate errors due to adjustment of blood and brain curves and peaks. Figure 3 shows a comparison of results applying different fit strategies to the same set of data.

Lumped constant

The lumped constant accounts for the differences in transport and phosphorylation. It has been estimated originally in humans as a normalization constant by comparing the dynamically determined CMR_{Glc} in whole brain of normal subjects assuming 50% gray and 50% white matter with the known value of whole brain glucose consumption (Phelps et al., 1979). This resulted in a value of LC = 0.42 which differs from the directly measured value of LC = 0.52 (Reivich et al., 1985). Both values are now widely used in different laboratories.

Instrumental inaccuracies

Beside the approximations used in the modeling to derive the glucose consumption rate from measured activity concentrations, the physical measurement of the tracer concentration itself is already afflicted by several inaccuracies due to the methods of PET.

Spatial resolution

With present day technology commercial tomographs have not yet reached the inherent physical limitations of the achievable spatial resolution which are due to positron physics. Therefore, it is determined by the camera design and varies between approximately 4 mm for the modern high resolution tomographs to greater than 10 mm for older instruments. There is also a variation of the resolution through the field of view i.e. through the image plane, which may differ between the planes (direct and cross planes). It also depends on the data acquisition mode, counting statistics and the parameters used for image reconstruction.

Attenuation correction

Since the annihilation photons have a high probability to interact with the surrounding tissue on their way to the detectors, huge attenuation corrections have to be applied to the measured data. Two methods are commonly used: 1) calculated attenuation corrections using either a simple geometric shape (an ellipse for the head) or information on the boundaries of the object from the analysis of short transmission measurements (Huang et al., 1981) or from the measured projection data (Bergström et al., 1982); 2) direct measurement of the absorption with a separate transmission scan. Both methods have their drawbacks. Transmission scans are time consuming, they add additional radiation dose and the subject may not change his position during the scans. The calculated corrections mostly neglect the influence of varying skull thickness and the attenuation of the head holder and the reconstructed pixel values may be very sensitive to misalignments of the head contour.

Scatter correction

PET systems especially designed for brain studies have a closer geometry and therefore a higher scatter to true events ratio in the order of 25%. This makes the correction for scatter more important than for whole body systems. Neglect of scatter corrections results in images with values that are erroneously too high. These errors are much larger in regions with low activity concentration, e.g. white matter regions. This causes additionally an overall reduction in contrast in the images.

Input function

For the quantification of images of regional activity concentrations in terms of CMR_{Glc} the time course of tracer activity in arterial blood the so-called "input function" is needed. It can either be measured by drawing manually or automatically arterial blood samples from a radial artery. To reduce trauma to the subject arterialized venous blood drawn from a hand heated to about 44°C may be used instead. An automated blood sampling system which draws and measures arterial blood continuously allows to correct for the shift in time and the dispersion effects between the measured blood activity time course and the input function at the locus of the brain. This may be of importance for the correct estimation of the model parameters with dynamic scanning.

Discussion

The FDG-model is applied in two different ways with PET:

Autoradiographic method or single scan method

Here a single measurement of the radioactivity distribution in brain tissue is made at around 40 minutes after injection of 5–10 mCi FDG. Cerebral metabolic rate for glucose (CMR_{Glc}) is calculated from an operational equation which needs in addition to the measured tissue activity the time course of the tracer activity in blood plasma from the time of injection until the time of measurement and average values of the model rate constants describing the tissue kinetics of FDG.

Dynamic method

Here sequential measurements of the time course of regional tracer accumulation in tissue are made over approximately 40 min. CMR_{Glc} is calculated from the detailed uptake curves and time course of the plasma activity yielding also the individual rate constants.

Both methods require a conversion factor between deoxyglucose and glucose the "lumped constant" (LC) which must be known for the brain regions of interest. This factor which is generally assumed to be regionally constant incorporates the different affinities of deoxyglucose and glucose for transport across the blood-brain barrier and phosphorylation by hexokinase. It has been shown that the LC may change in abnormal states, but it is common practice to assume a constant value for LC which, however, may differ between groups. As long as the value used for LC is known, results from different groups can easily be converted.

Other differences between groups in performing and analyzing FDG-studies cannot so easily be taken into account:

Average values of the rate constants in the autoradiographic model
The use of average normal values for the rate constants may lead to erroneous results especially in pathologic states with largely reduced CMR_{Glc}. For the normal range of CMR_{Glc} the values do not depend strongly on those average rate constants. However, an abnormally high glucose level in plasma may also result in largely reduced values for the model rate constant which can not be approximated by the use of normal values in Eq. 4. Therefore, the use of a modified operational equation (Brooks, 1982; Hutchins et al., 1984; Rhodes et al., 1983; Wienhard et al., 1985) is recommended, which is less sensitive to deviations of the subject rate constants from the population mean rate constants. It should be kept in mind that the various equations have different error sensitivities. If a modified operational equation is used it should clearly be stated.

Including a k_4 in the model
The Sokoloff model has been extended to take glucose phosphatase activity into account in the model (Phelps et al., 1979). For FDG-studies not exceeding scan times beyond 50 min after injection it does not seem necessary to include a k_4 in the model. However, assuming $k_4 = 0$ results in CMR_{Glc} values which are different from those obtained by allowing k_4 to vary or using a fixed k_4 (Lammertsma et al., 1987).

Time protocol of the PET-study
For the autoradiographic model, scans accumulating data between 30 and 60 min after injection seem to be appropriate. In dynamic studies short scan times (<60 sec) during the first few minutes are necessary, especially with rapid bolus injections and if blood volume contributions to issue activity are included in the model. Dynamic scan time should at least total to 30 min. If k_4 is included in the model and is to be determined from fitting to the dynamic data, then even 60 min scanning time is much too short to get realistic and reliable k_4 values (Lammertsma et al., 1987).

Blood sampling
For the autoradiographic model manual sampling of arterialized venous blood seems adequate; the proper arterialization should be checked by

measuring pO_2. For a dynamic analysis automated sampling of arterial blood and corrections for time shift and dispersion are the optimal conditions if individual rate constants are to be determined. If only CMR_{Glc} is calculated from dynamic scans, arterialized venous blood sampled manually seems sufficient.

Inclusion of a blood volume term in the model

This is only necessary if individual rate constants are to be determined from dynamic scans (Lammertsma et al., 1987). An alternate pragmatic approach is to discard the data during the first few minutes after injection (Wienhard et al., 1985), or to use a nominal value for regional cerebral blood volume (Phelps et al., 1979; Huang et al., 1980). This, of course, is less accurate than measuring it individually with a $C^{15}O$-inhalation study before performing the FDG scan.

Administration of the tracer: rapid bolus or slow infusion

The way of tracer administration also is only important if individual rate constants are to be determined. A slow infusion is less sensitive to timing and dispersion. A rapid bolus gives better defined conditions if one wants to correct for those effects and shows less sensitivity to errors due to tissue heterogeneity (Herholz et al., 1987). From a practical computational point of view, with a rapid bolus injection the plasma tracer time course can be described as decaying exponential functions thus allowing to solve the model equations analytically with no need for lengthy numerical integrations of the model differential equations.

Stable plasma glucose level

For the model analysis a stable plasma glucose level over the study period is necessary. Fasting of the patient for several hours before the investigation may help to achieve it. The plasma glucose level should be checked at several times over the study period.

Purity of the tracer

Many studies have been performed with FDG contaminated with various degrees of fluorodeoxymannose (FDM). If the amount of FDM is known, it can be corrected for by a modified value of the lumped constant (Wienhard et al., 1991), since no significant regional variability of the differential tracer behavior was observed in normal or in lesioned brain tissue.

External environment

A reproducible external environment is important to avoid unwanted stimulation effects. Many groups prefer low light and low ambient noise with eyes and ears open.

Anxiety of the patient

Anxiety of the patient may alter CMR_{Glc}-values. Therefore, one should try to make the patient familiar with the whole procedure to keep these effects as low as possible.

Use of a head fixation device

The use of a head fixation device may help to directly relate CT and MR images and also prevent patients head movements (Bergström et al., 1981).

Calculated or measured attenuation correction

Measured attenuation corrections are more accurate and are important if one looks for small effects in CMR_{Glc}. Practical reasons may make it necessary to compromise on this (extended study duration, the need to shift the patient because of limited axial field of view with elder tomographs) and adjust an ellipse to the brain contour.

Misalignment of the ellipse may in a clinical situation be misinterpreted as an asymmetry in the tracer uptake. This can be improved by taking the brain border contour from the measured projection data. With some effort this method can be improved to include also corrections for bone and head holder (Michel et al., 1989).

Corrections for scattered radiation

Neglecting proper corrections for scatter contributions causes large errors in CMR_{Glc} and loss of image contrast, especially with dedicated brain scanners. If the object is not centered and symmetrical in the field of view of the tomograph, neglect of scatter corrections may cause artificial asymmetries in the reconstructed images.

Instrumental spatial resolution

The spatial resolution of the tomograph determines the degree of partial volume effects, recovery coefficients, gray/white matter ratios etc. Therefore, the system resolution is an important parameter which should always be quoted.

Definition of regions of interest

The definition of regions of interest is a topic of numerous discussions and many different approaches are in use. There is considerable need for automation and standardization. In any case as much information as possible from other imaging devices like CT and MR should be used together with the PET information.

All these issues influence CMR_{Glc}-values to some extent and only few of them can be standardized or properly taken into account when results of different groups are compared. Therefore, as much information as possible should be given how all these topics were handled during the performance and the analysis of a PET study.

References

Baron JC, Rougemont D, Soussaline F, Bustany P, Crouzel C, Bousser MG, Comar D (1984) Local interrelationships of cerebral oxygen consumption and glucose utilization in normal subjects and in ischemic stroke patients: a positron tomography study. J Cereb Blood Flow Metab 4: 140–149

Bergström M, Boethius J, Eriksson L, Greitz T, Ribbe T, Widén L (1981) Head fixation device for reproducible position alignment in transmission CT and positron emission tomography. J Comput Assist Tomogr 5: 136–141

Bergström M, Litton J, Eriksson L, Bohm C, Blomqvist G (1982) Determination of object contour from projections for attenuation correction in cranial positron emission tomography. J Comput Assist Tomogr 6: 365–372

Brooks RA (1982) Alternative formula for glucose utilization using labelled deoxyglucose. J Nucl Med 23: 528–539

Gjedde A, Wienhard K, Heiss WD, Kloster G, Diemer NH, Herholz K, Pawlik G (1985) Comparative regional analysis of 2-fluorodeoxyglucose and methylglucose uptake in brain of four stroke patients. With special reference to the regional estimation of the lumped constant. J Cereb Blood Flow Metab 5: 163–178

Hawkins RA, Phelps ME, Huang SC, Kuhl DE (1981) Effect of ischemia on quantification of local cerebral glucose metabolic rate in man. J Cereb Blood Flow Metab 1: 37–51

Hawkins RA, Phelps ME, Huang SC (1986) Effects of temporal sampling, glucose metabolic rates and disruptions of the blood brain barrier on the FDG model with and without a vascular compartment: studies in human brain tumours with PET. J Cereb Blood Flow Metab 6: 170–183

Heiss WD, Wienhard K, Pawlik G, Wagner R, Ilsen HW, Herholz K (1983) Hypometabolism in stroke: cerebral metabolic rate for glucose in infarcted and remote tissue obtained by dynamic determination of individual kinetic constants. In: Greitz T, Ingvar DH, Widén L (eds) The metabolism of the human brain studied with positron emission tomography. Raven Press, New York, pp 399–409

Herholz K (1988) Non-stationary spatial filtering and accelerated curve fitting for parametric imaging with dynamic PET. Eur J Nucl Med 14: 477–484

Herholz K, Patlak CS (1987) The influence of tissue heterogeneity on results of fitting non-linear model equations to regional tracer uptake curves: with an application to compartmental models used in positron emission tomography. J Cereb Blood Flow Metab 7: 214–229

Huang SC, Phelps ME, Hoffman EJ, Sideris K, Selin CJ, Kuhl DE (1980) Non-invasive determination of local cerebral metabolic rate of glucose in man. Am J Physiol 238: 569–582

Huang SC, Carson RE, Phelps ME, Hoffman EJ, Schelbert HR, Kuhl DE (1981) A boundary method for attenuation correction in positron computed tomography. J Nucl Med 22: 627–637

Hutchins GD, Holden JE, Koeppe RA, Halama JR, Gatley SJ, Nickles RJ (1984) Alternative approaches to single scan estimation of cerebral glucose metabolic rate using glucose analogs, with particular application to ischemia. J Cereb Blood Flow Metab 4: 35–40

Lammertsma AA, Brooks DJ, Frackowiak RSJ, Beaney RP, Herold S, Heather JD, Palmer AJ, Jones T (1987) Measurement of glucose utilization with (^{18}F)2-fluoro-2-deoxy-D-glucose: a comparison of different analytical methods. J Cereb Blood Flow Metab 7: 161–172

Mazoyer BM, Huesman RH, Budinger TF, Knittel BL (1986) Dynamic PET data analysis. J Comput Assist Tomogr 10: 645–653

Michel C, Bol A, De Volder AG, Goffinet AM (1989) Online brain attenuation correction in PET: towards a fully automated data handling in a clinical environment. Eur J Nucl Med 15: 712–718

Patlak CS, Blasberg RG, Fenstermacher JD (1983) Graphical evaluation of blood-to-brain transfer constants from multiple-time uptake data. J Cereb Blood Flow Metab 3: 1–7

Phelps ME, Huang SC, Hoffman EJ, Selin C, Sokoloff L, Kuhl DE (1979) Tomographic measurement of local cerebral glucose metabolic rate in humans with (F-18)2-fluoro-2-deoxy-D-glucose: validation of the model. Ann Neurol 6: 371–388

Reivich M, Kuhl D, Wolf A, Greenberg J, Phelps M, Ido T, Casella V, Fowler J, Hoffman E, Alavi A, Sokoloff L (1979) The ^{18}F-fluorodeoxyglucose method for the measurement of local cerebral glucose utilization in man. Circ Res 44: 127–137

Reivich M, Alavi A, Wolf A, Fowler J, Arnett C, MacGregor RR, Shiue CY, Atkins H, Anand A, Dann R, Greenberg JH (1985) Glucose metabolic rate kinetic model parameter determination in humans: the lumped constants and rate constants for ^{18}F-fluorodeoxyglucose and ^{11}C-deoxyglucose. J Cereb Blood Flow Metab 5: 179–192

Rhodes CG, Wise RJS, Gibbs JM, Frackowiak RSJ, Hatazawa J, Palmer AJ, Thomas DGT, Jones T (1983) In vivo disturbance of the oxidative metabolism of glucose in human cerebral gliomas. Ann Neurol 14: 614–626

Sasaki H, Kanno I, Murakami M, Shishido F, Uemura K (1986) Tomographic mapping of kinetic rate constants in the fluorodeoxyglucose model using dynamic positron emission tomography. J Cereb Blood Flow Metab 6: 447–454

Sokoloff L, Reivich M, Kennedy C, DesRosiers MH, Patlak CS, Pettigrew KD, Sakurada O, Shinohara M (1977) The ^{14}C-deoxyglucose method for the measurement of local cerebral glucose utilisation: theory, procedure and normal values in the conscious and anaesthetized albino rat. J Neurochem 28: 897–916

Wienhard K, Pawlik G, Herholz K, Wagner R, Heiss WD (1985) Estimation of local cerebral glucose utilisation by positron emission tomography of ^{18}F-2-fluoro-2-deoxy-D-glucose: a critical appraisal of optimisation procedures. J Cereb Blood Flow Metab 5: 115–125

Wienhard K, Pawlik G, Nebeling B, Rudolf J, Fink G, Hamacher K, Stöcklin G, Heiss WD (1991) Estimation of local cerebral glucose utilisation by positron emission tomography: comparison of (^{18}F)-2-fluoro-2-deoxy-D-glucose and (^{18}F)-2-fluoro-2-deoxy-D-mannose in patients with focal brain lesions. J Cereb Blood Flow Metab 11: 485–491

Author's address: Prof. Dr. K. Wienhard, Max-Planck-Institut für Neurologische Forschung, Gleueler Strasse 50, D-W-5000 Köln 41 (Lindenthal), Federal Republic of Germany

J Neural Transm (1992) [Suppl] 37: 53–66

On the influence of spatial resolution and of the size and form of regions of interest on the measurement of regional cerebral metabolic rates by positron emission tomography

T. Kuwert[1], T. Sures[1], H. Herzog[1], M. Loken[1,2,*], M. Hennerici[3], K.-J. Langen[1], and L. E. Feinendegen[1]

[1] Institute of Medicine, Research Center Jülich, Federal Republic of Germany
[2] Division of Nuclear Medicine, University of Minnesota Medical Center, Minneapoli, USA (Emeritus)
[3] Department of Neurology, University of Heidelberg, Klinikum Mannheim, Federal Republic of Germany

Summary. Factors that affect the accuracy of the positron emission tomographic (PET) quantification of cerebral metabolic rates include the spatial resolution of the employed imaging device and the method used for extraction of regional metabolic values from the PET data set. The present article reviews (i) how and to what extent these two factors are presumed to influence the measurement of absolute values of cerebral metabolic rates and their ratios, and (ii) whether and how these factors may affect comparisons of regional metabolic rates between groups of subjects.

Introduction

The extraction of regional cerebral metabolic data from images generated by positron emission tomography (PET) has been the subject of several publications (Mazziotta et al., 1981, 1987, 1991; Herholz et al., 1985; Bohm et al., 1983, 1986, this volume; McNamara et al., 1987; Evans et al., 1988a,b, 1989, 1991; Fox and Kall, 1987; Fox, 1991; Levy et al., 1991; Marrett et al., 1989; Rottenberg et al., 1991; Valentino et al., 1988; Seitz et al., 1990). Although some methods have been devised which use information from all picture elements (pixels) of the image (e.g., Fox and Kall, 1987; Fox, 1991; Levy et al., 1991), most current strategies require the selection of pixel subsets related to cerebral anatomy, the so-called regions of interest (ROIs). Metabolic values are then calculated as the average of all pixel values in a given region and thought to give an estimate of a biochemical function related to a specific region of the brain.

---

* Alexander von Humboldt Awardee 1990-91

Factors affecting the accuracy of this measurement include difficulties in localizing the ROI with relation to cerebral anatomy, as reviewed by Bohm et al. (this volume), and the relation of spatial resolution and the size and the form of the ROIs used to the size and the form of the cerebral structures evaluated.

In PET, metabolic values obtained in an individual patient are either compared to those obtained in a group of normal subjects for diagnostic purposes or grouped together with those obtained in other patients in order to allow comparisons between groups of subjects for research purposes. Therefore, it is an important aspect whether and how the comparison between metabolic values measured in different subjects or different groups of subjects is affected by such parameters of PET technology as image resolution or method of regionalisation.

The present article will therefore first consider how image resolution and ROI size and form might influence the accuracy of single measurements, and then briefly give an account of the effect of these parameters on comparisons of metabolic values measured in different groups of subjects.

The influence of spatial resolution

To date, despite considerable advances in scanner technology and especially in image resolution during the last ten years, the majority of absolute metabolic values reported for structures of gray matter in PET studies can still be considered underestimations of the real values. This underestimation is due to the partial volume effect (Hoffman et al., 1979) occurring because the size of most cerebral structures of interest is smaller than twice the optimal spatial resolution of present PET cameras, so that ROI measurements invariably include measurements of areas adjacent to the ROI considered. Equality in size to the double of resolution, however, would be necessary for accurate quantification of the isotope concentration in the tissue as shown by Hoffman et al. (1979), who found in a phantom study that a cylindrical structure equal in width to the resolution of the imaging system used will have only 50% of the correct isotope concentration in the image.

Partial volume effects may act either axially, that is, in the z-axis of the slices recorded, or transaxially, that is, in the plane of the image.

Axial partial volume effects are dependent on the spacing and the thickness of the image planes relative to the size and position of the cerebral structure evaluated (Kearfott and Kluksdahl, 1989). The axial resolution is primarily a function of the slice thickness of the individual image planes, approaching 20 mm in older PET cameras and approximately 5 mm for recent multislice cameras with a block detector design (Eriksson et al., 1986; Murayama et al., 1982; Nutt et al., 1985). Within a given plane the z-resolution is dependent on the radial distance from the centre due to the fact that detector rings are inhomogeneously sensitive in the z-direction

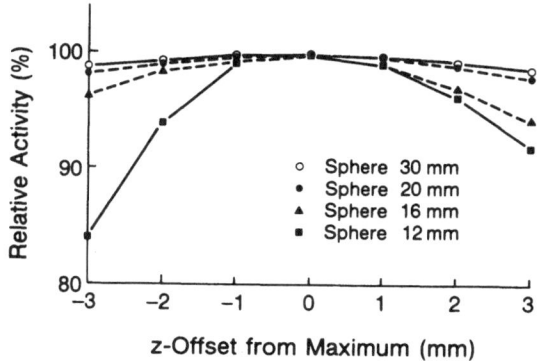

Fig. 1. Recovery of concentration of radioactivity (^{18}F) instilled in spheres of different diameters expressed as a function of the offset in z — direction from the middle of the slice

leading to variations of 80% to 100% of the recorded value when a point source is being moved along the z-axis (Rota Kops et al., 1990). Figure 1 illustrates the magnitude of this effect for volumes larger than a point source and the PC-4096 PET camera; depending on the size of spheres filled with a radioactive fluid the underestimation of the real value may amount up to 15% if a sphere with a diameter of 12 mm is used, and may still reach 4% when the diameter of the sphere is 16 mm (Herzog et al., 1991) and thus nearly three times the axial spatial resolution of the camera used.

Although some approaches are being developed to quantify metabolic values three-dimensionally, either by specially designed cameras (Karp et al., 1991) or by using software designed to yield exactly matching magnetic resonance imaging (MRI) overlays (Pelizarri et al., 1989; Evans et al., 1989, 1991; Mazziotta et al., 1991), these facilities are at the present time confined to comparatively few clinical centers. Thus regional analysis relies largely on the examination of two-dimensional brain slices warranting a separate discussion of transaxial partial volume effects caused by the limited transaxial resolution of the image.

Using a digitized neuroanatomical brain slice simulating a PET image of regional cerebral glucose utilization, Mazziotta et al. (1981) have determined that — if an imaging device with a transaxial spatial resolution of 5 mm, i.e., the state-of-the-art performance of to date PET systems was used — the underestimation of the true value ranged from approximately 5 to 25%, depending on the cerebral structure evaluated. The errors were largest for small, thin, and irregularly shaped structures, whose values were most different from those in adjacent structures, and smallest for large, circular structures surrounded by regions with similar values.

Very few studies have systematically related the dependence of recorded metabolic values in the living human brain to the resolution of the imaging system used. Grady et al. (1989) studied a group of subjects with two different scanners. They reported that measures of glucose utilization

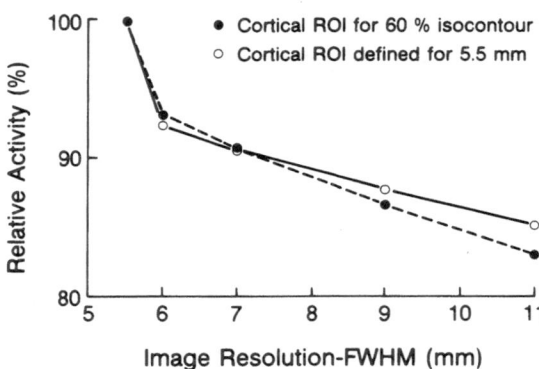

Fig. 2. The dependence of FDG uptake measured in cortical regions expressed as a function of image resolution. The uptake found at the different spatial resolutions is related to the uptake measured for an image resolution of 5.5 mm (relative activity)

ranged from 30 to 120% higher when the scanner PC1024 with an image resolution of 6 mm full-width half-maximum (FWHM) was used than those obtained by the ECAT II with an intrinsic resolution of 17 mm FWHM. The per cent increase was the greatest in those regions with the lowest values from the ECAT II, which were assumed to be the ones most under-estimated due to the partial volume effect in the ECAT II measurements.

The differences in glucose utilization between these two scanners, however, cannot be attributed exclusively to differences in spatial resolution, since the two cameras differ also in other technical factors such as the scatter fraction and since two different methods of ROI definition were used — the ECAT II values were obtained by drawing comparatively large ROIs by hand, whereas the PC1024 values were measured in relatively small circles with a diameter of 8 mm, minimizing partial volume artifacts.

Data obtained by reconstruction of images depicting the cerebral accumulation of ^{18}F-fluorodeoxyglucose (FDG) in a normal subject using five different cut-off frequencies, thus creating five images with different in-plane resolution (Herzog et al., 1991; Fig. 2), indicate that the dependence of PET values on image resolution is non-linear. Relative cortical activity, i.e., the ratio between cortical activity recovered in the PET image with the highest resolution and cortical activity measured in those with lower resolution, decreased rather sharply by 7% comparing values obtained at a resolution of 5.5 mm to those for an image resolution of 5.8 mm. A further decrease of image resolution from 5.8 mm to 11 mm caused only an additional loss of 8% in the accuracy of quantification. In this study, cortical ROIs were defined using isocontours at a level of 60% of the ROI's maximum, simulating manual definition of ROIs directly on the PET image, which is a widely used method of regionalisation (Kuwert et al., 1990; Goffinet et al., 1989).

Since absolute metabolic values obtained by PET vary greatly between individuals (Kuhl et al., 1982; Tyler et al., 1988), attempts to reduce

Table 1. Ratios of regional cerebral glucose consumption determined in nine patients with unilateral thalamic infarction: dependence on spatial resolution

Ratio	Image resolution (FWHM/mm)[3]		
	7.1	8.9	11
Cortex A/U[1,2]	0.87	0.88	0.89
Thalamus A/U	0.81	0.80	0.79
Caudate nucleus A/U	0.79	0.79	0.77
Thalamus A/Cortex U	1.03	0.87	0.82
Caudate A/Cortex U	0.79	0.58	0.52

[1] Regional cerebral metabolic rate measured in the hemisphere affected (A) by the infarction divided by metabolic rate measured in the unaffected hemisphere (U)
[2] Prefrontal cortex
[3] Images of different spatial resolution obtained by reconstruction of the same set of raw data using three different cut-off frequencies for the filtered backprojection

this variability by calculating ratios between metabolic values in different regions are frequently made in clinical PET studies. Despite the wide use of these ratios, only few studies have analysed systematically their dependence on image resolution.

Grady et al. (1989, 1991) reported that ratios between lobar glucose utilization and mean gray glucose utilization were relatively unaffected by differences in image resolution. Most probably the reason for the insensitivity of these ratios to alterations of resolution is that numerator and denominator are affected to the same extent by partial volume artifacts. This is illustrated by the observation that ratios calculated between homologous ROIs are relatively independent on image resolution, whereas this is not true of ratios calculated between different areas of the brain (see Table 1; authors' unpublished data).

This also becomes apparent when the ratio between glucose consumption measured in gray matter and in white matter is considered. This ratio, which was autoradiographically determined in the monkey to approximate 4:1 (Kennedy et al., 1978), was less than 2:1 when older PET cameras with an in-plane resolution of approximately 17 mm were used. It rose to a value of 3.1:1 in images generated by new-generation scanners such as the PC4096 (Rota Kops et al., 1990).

In summary, we conclude that both absolute and relative metabolic values determined by PET are largely dependent on the image resolution of the PET camera used. This dependence seems to be non-linear and varies considerably among different brain areas, making it difficult to compare values measured by PET cameras with different spatial resolution and precluding the comparison of absolute values between studies using different imaging devices.

The effect of ROI size and form

The partial volume effect described above also affects measurements in single pixels which thus exhibit metabolic values integrating the value related to the area directly underlying this pixel with those of its surrounding structures in a rather complex fashion. Due to the limited spatial resolution of the imaging devices at hand, it is not possible to select subsets of pixels reflecting metabolism exclusively in one specific region of the brain. To date, ROI values can therefore only be considered to be correlated to the true biochemical variable, but do not solely reflect metabolism in that structure.

Besides nonanatomical statistical approaches to regionalization (e.g., Fox and Kall, 1987; Fox et al., 1988; Mintun et al., 1989; Fox, 1991; Lueck et al., 1989; Levy et al., 1991) two principal ways of defining ROIs can be distinguished: first, the definition of ROIs on structural images, that is on CT or MRI images, and secondly, their definition on the PET image itself.

The first method, which is described elsewhere in this volume is indispensable when small structures in the brain, especially in the cortex, are to be localised. Although ROI values generated by one of these methods are also affected by the partial volume effect for reasons described above, the delineation of ROIs using structural images of the brain poses considerably less problems due to the better spatial resolution of structural imaging techniques.

This is not the case when the ROIs are directly defined on the PET images, since the boundaries of anatomical structures seem blurred by partial volume artifacts. Generally, three approaches seem feasible: definition of irregular ROIs by visual extraction either manually or by using isocontours; definition of geometric ROIs, for example boxes, ellipses, or circles; or methods relying on histographical analysis of the image, which can be fully automated if, for example, boundary-finding algorithms are used, or manual if, for example, maximal values of certain structures are selected. Some published methods may represent combinations of these possibilities.

Definition of irregular ROIs is a widely used method (see e.g., Goffinet et al., 1989; Kuwert et al., 1990) and offers the advantage of creating ROIs that closely delineate the often irregular boundaries of cerebral structures of interest compensating for individual differences in brain size and minimizing partial volume artefacts. If visual extraction is used, special care should be taken to avoid systematic subjective errors, including evaluation independent of knowledge on the subjects' individual diagnosis, standardization of range-setting on the screen, and the performance of the joint analysis of normal subjects and patients consecutively over a short period of time, and,

when possible, by the same observer. Also, an analysis of correlation
between ROI size and metabolic value to prove the independence of the
PET value on ROI size and a comparison between ROI sizes measured in
different groups of subjects to test for the homogeneity of the analysis (e.g.,
Kuwert et al., 1990) might be useful.

A semiautomated approach to defining irregular ROIs selects those
pixels in a larger manually delineated ROI that fall within a predetermined
range specified by lower and upper boundary values, the pixel with the
highest value in the manually defined ROI being selected as the upper
boundary (Rottenberg et al., 1991). The selection of the lower boundary is
usually arbitrary, different percent values leading to differences in ROI size
and thus to the extraction of different metabolic values, which is most
pronounced in high-resolution images (Rottenberg et al., 1991).

Numerous clinical PET studies advocate the use of geometric ROIs for
regionalisation of PET studies, since these may be placed tangentially to
each other and to an outer isocontour of the cortex or centered on the
maximal values for subcortical structures and minimize observer bias (e.g.,
Laplane et al., 1989; Martinot et al., 1990; Kuwert et al., 1991). Usually,
the size of these ROIs is arbitrarily chosen and constitutes a compromise
reflecting the distribution of cortical width as it appears in the PET image so
that regionally different partial volume effects occur as a function of the
different width and size of the structures evaluated. The most often used
templates are circles ranging from 8 to 40 mm in diameter. Limited knowledge exists concerning the influence of ROI size on the height of the PET
parameter investigated. The authors' data indicate that this influence may
be important and that metabolic values measured in the caudate nucleus
may decrease by nearly 40% if the ROI diameter increases from 1 to 2 cm
(Kuwert et al., 1991, submitted for publication). This depends, however, on
image resolution in the sense that changes in ROI values caused by enlarge-
ment in ROI size are greatest in the high-resolution images (Fig. 3).

Several authors use boundary-finding algorithms in their ROI schemes,
mostly to define the outer contour of the brain. This contour is used either
to align circular ROIs or to constitute the outer boundary of ROIs, their
inner boundary then being defined by a constant distance from the outer
contour (e.g., Buchsbaum et al., 1984). Given adequate software, this
analysis can be performed rapidly and minimizes observer bias. Here again,
ROI size is arbitrarily chosen. To our knowledge, no analysis of the effect
of ROI size on ROI values has been published.

The analysis of histograms has been introduced as a means of ROI
identification and of ROI delineation proving particularly useful in conditions producing gross atrophy of brain structures of interest, as in the case of
the caudate nucleus in manifest Huntington's disease (Kuhl et al., 1982;
Young et al., 1986; Kuwert et al., 1990). Strategies relying on this approach
identify peak values of the structure of interest by analyzing cross-sectional
histograms, thus minimizing, but not completely avoiding, the partial
volume effect. An analogous technique for the measurement of cortical

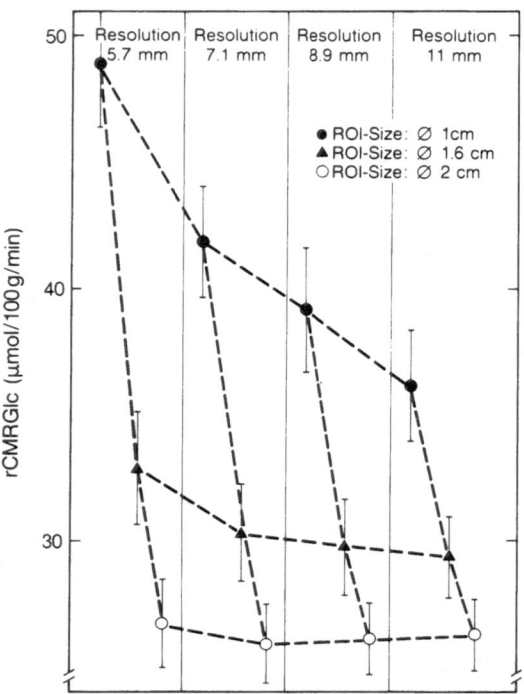

Fig. 3. The dependence of caudate glucose consumption on the size of circular ROIs and transaxial spatial resolution as determined in eleven normal subjects. The bars indicate errors of the means. The images with different spatial resolution have been generated by a variation of the cutoff frequency used for the filtered backprojection

metabolism using cross-sectional histograms has been published (McNamara et al., 1987), but seems to be used only in a minority of PET studies.

In summary, we conclude that numerous approaches to regionalization of PET data are in current use and that only limited knowledge exists concerning the dependence of ROI values on the characteristics of the different ROI schemes. As expected, there seems to be a close relationship between ROI size and the ROI metabolic values, due to the partial volume effect. This relationship becomes increasingly important with an improvement in spatial resolution of the imaging device.

The effect of image resolution and ROI size on the significance of comparisons between metabolic values measured in groups of subjects

The two preceding chapters of this article have addressed how image resolution and how the size and the form of ROIs are supposed to influence the accuracy of metabolic measures determined by PET. It was concluded that — due to partial volume artifacts — present measurements of metabolic values in gray matter represent underestimations of the real value, depending strongly on the limitations of image resolution and the method of regionalization.

To date and especially in psychiatry, PET is mainly a research tool. The outcome of clinical PET studies, however, are not single values measured in one subject, but, rather, levels of significance describing differences in metabolic values between groups of subjects. An important aspect of PET technology therefore is not only its accuracy, but also how its different parameters influence the outcome of group comparisons of metabolic values.

Although this aspect is crucial for comparing results obtained by studies using different PET technology and for study design, only little evidence exists concerning this relationship so that only a brief and preliminary account on this subject can be given here.

One frequently made assumption is that small differences in metabolic values between groups require the use of high-resolution PET scanners together with small ROIs minimizing partial volume artifacts. This assumption, however, may not be generally true as indicated by the following evidence (authors' own unpublished observation).

In eleven normal subjects and nine patients with unilateral thalamic infarctions, images of cerebral glucose utilization measured by the PET scanner PC4096 were reconstructed using four different cut-off frequencies for the filtered back-projection generating images with four different image resolutions ranging from 5.7 to 11 mm. Prefrontal glucose consumption, which may be reduced in patients ipsilateral to thalamic infarction due to deafferentation (Kuwert et al., 1991), was then evaluated using sets of circular ROIs of three different diameters ranging from 1 to 2 cm. Subsequent statistical analysis using Kruskal-Wallis tests (see Tables 2 and 3) revealed that, for the comparison of group means of frontal glucose consumption or its bilateral ratios, the levels of significance attained were all below 0.03 regardless of the combination of ROI size and image resolution used, confirming results obtained for caudate glucose consumption using the same approach (Kuwert et al., 1991, submitted for publication). Whereas an increase in ROI size led to a worsening of the levels of significance

Table 2. Levels of significance determined using Kruskal-Wallis tests assessing differences in prefrontal glucose consumption between patients and controls

Ø ROI	IR = 5.7 mm	IR = 7.1 mm	IR = 8.9 mm	IR = 11 mm
1.0 cm	0.0070	0.0070	0.0088	0.0088
1.6 cm	0.0167	0.0205	0.0167	0.0167
2.0 cm	0.0205	0.0250	0.0250	0.0205

IR Image resolution. Differences in IR were generated by varying the cut-off frequency used for the filtered back projection
Ø ROI ROI size is given as the diameter of the circular ROI used
All values are levels of significance

Table 3. Levels of significance determined using Kruskal-Wallis tests assessing differences in bilateral ratios of prefrontal glucose consumption between patients and controls

Ø ROI	IR = 5.7 mm	IR = 7.1 mm	IR = 8.9 mm	IR = 11 mm
1.0 cm	0.0007	0.0016	0.0007	0.0004
1.6 cm	0.0021	0.0012	0.0016	0.0027
2.0 cm	0.0034	0.0044	0.0027	0.0016

IR Image resolution. Differences in IR were generated by varying the cut-off frequency used for the filtered back projection
Ø ROI ROI size is given as the diameter of the circular ROI used
All values are levels of significance

attained, no major effect of image resolution can be observed (Tables 2 and 3). The latter finding contrasts with the results obtained for caudate glucose consumption (Kuwert et al., 1991, submitted for publication) where a worsening of image resolution led to an improvement of the levels of significance assessing differences between patients and controls due to an increased variability of the metabolic values encountered at high image resolutions. This increased variability was believed to stem from an increase in image noise on the high resolving images generated by the use of low cut-off frequencies for the filtered backprojection (Rota Kops et al., 1990). In the evaluation of prefrontal glucose consumption, this effect might have been obscured since prefrontal glucose consumption was measured as the average metabolic value measured in several, i.e. (depending on ROI size) three to seven circular ROIs, thus reducing variability, whereas caudate glucose metabolic rates were evaluated in only one ROI.

Analyses of the effect of methodological parameters on the outcome of group comparisons have also been reported by Rottenberg et al. (1991) and Strother et al. (1991) who used thresholding (see above) for regionalisation and a factor analytic approach, the so-called scaled subprofile model (SSM; Moeller et al., 1987), for statistical analysis of cerebral glucose consumption measured in normal subjects and patients with AIDS dementia. They have shown that SSM-derived covariance patterns are independent of the selection of different lower boundaries leading to the creation of ROIs of different sizes and of the optimization of various parameters of image reconstruction including image resolution, transmission scan smoothing, and the application of a scatter deconvolution correction. One of their conclusions is that the insensitivity of their technique to discriminate different groups of subjects to variations of the above-mentioned parameters was due to the application of the SSM analysis, which, however, might not necessarily represent the only interpretation in view of the data presented above (Kuwert et al., 1991, submitted for publication), obtained by simpler statistical techniques.

Although the three above-mentioned studies agree in the observation that — contrary to the metabolic values themselves — levels of significance describing group differences of metabolic values are relatively insensitive to variations of spatial resolution or ROI size, caution should be exerted in generalizing this observation in view of the complexity of PET methodology. Further systematic analyses of the relationship between parameters of PET technology and the detection of statistically significant group differences in cerebral metabolic rates are necessary to clarify this relationship. This is relevant to both study design and to the interpretation of diverging results reported in PET studies using different methodologies. In further studies special attention should be given not only to the influence of technical parameters on group means of metabolic values, but also to their influence on the variation of these values within groups of subjects.

Acknowledgements

The authors gratefully acknowledge the thorough revision of this manuscript by Prof. K.I. Altman and the secretarial help of M.D. Beaujean.

References

Bohm C, Greitz T, Kingsley D, Berggren B, Ollson L (1983) Adjustable computerized stereotaxic brain atlas for transmission and emission tomography. AJNR 4: 731–733

Bohm C, Greitz T, Blomquist G, Farde L, Forssgren PO, Kingsley D, Sjögren I, Wiesel FA, Wik G (1986) Applications of a computerized adjustable brain atlas in positron emission tomography. Acta Radiol [Suppl] 369: 449–452

Buchsbaum MS, Cappelletti J, Ball R, Hazlett E, King AC, Johnson J, Wu J, DeLisi LE (1984) Positron emission tomographic image measurement in schizophrenia and affective disorders. Ann Neurol 15 [Suppl]: S157–S165

Eriksson L, Bergström M, Bohm C, Holte S, Kesselberg M, Litton J (1986) Figures of merit for different detector configurations utilized in high resolution positron cameras. IEEE Trans Nucl Sci NS 33: 446–451

Evans AC, Beil C, Marrett S, Thompson CJ, Hakin A (1988a) Anatomical-functional correlation using an adjustable MRI-based region of interest atlas with positron emission tomography. J Cereb Blood Flow Metab 8: 513–530

Evans AC, Beil C, Marrett S, Thompson CJ, Hakim A (1988b) Anatomical functional correlation using an adjustable MRI based atlas with PET. J Cereb Blood Flow Metab 8: 813–830

Evans AC, Marrett S, Collins L, Peters TM (1989) Anatomical-functional correlative analysis of the human brain using three-dimensional imaging systems. Proc SPIE 1092:264–274

Evans AC, Marrett S, Torrescorzo J, Ku S, Collins L (1991) MRI-PET correlation in three dimensions using a volume-of-interest (VOI) atlas. J Cereb Blood Flow Metab 11: A69–A78

Fox PT (1991) Physiological ROI definition by image subtraction. J Cereb Blood Flow Metab 11: A79–A82

Fox PT, Kall B (1987) Stereotaxy as a means of anatomical localization in physiological brain images: proposals for further validation. J Cereb Blood Flow Metab 7: S18–S20

Fox PT, Mintun MA (1989) Noninvasive functional brain mapping by change-distribution analysis of averaged PET images of $H_2^{15}O$ tissue activity. J Nucl Med 30: 141–149

Fox PT, Mintun MA, Reiman E, Raichle ME (1988) Enhanced detection of focal brain responses using intersubject averaging and change-distribution analysis of subtracted PET images. J Cereb Blood Flow Metab 8: 642–653

Goffinet AM, De Volder AG, Gillain C, Rectem D, Bol A, Michel C, Cogneau M, Labar D, Laterre C (1989) Positron tomography demonstrates frontal lobe hypometabolism in progressive supranuclear palsy. Ann Neurol 25: 131–139

Grady CL (1991) Quantitative comparison of measurements of cerebral glucose metabolic rate made with two positron cameras. J Cereb Blood Flow Metab 11: A57–A63

Grady CL, Berg G, Carson RE, Daube-Witherspoon ME, Friedland RP, Rapoport SI (1989) Quantitative comparison of cerebral glucose metabolic rates from two positron emission tomographs. J Nucl Med 30: 1386–1392

Herholz K, Pawlik G, Wienhard K, Heiss W-D (1985) Computer assisted mapping in quantitative analysis of cerebral positron emission tomograms. J Comput Assist Tomogr 9: 154–161

Herzog H, Rota Kops E, Schmid A, Feinendegen LE (1991) A consideration of the effects of differing design parameters in PET system on the accuracy of radioactivity quantitation in vivo. Med Progr Techn (in press)

Hoffman EJ, Huang S-C, Phelps ME (1979) Quantitation in positron emission computed tomography. 1. Effect of object size. J Comput Assist Tomogr 3: 299–308

Karp JS, Daube-Whitherspoon ME, Muellehner G (1991) Factors affecting accuracy and precision in PET volume imaging. J Cereb Blood Flow Metab 11: A38–A44

Kearfott KJ, Kluksdahl EM (1989) Effects of axial spatial resolution and sampling on object detectability and contrast for multiplanar positron emission tomography. Med Phys 16: 785–790

Kennedy C, Sakurada O, Shinohara M, Jehle J, Sokoloff L (1978) Local cerebral glucose utilization in the normal conscious macaque monkey. Ann Neurol 4: 293–301

Kessler RM, Ellis JR, Eden M (1984) Analysis of emission tomographic scan data: limitations imposed by resolution and background. J Comput Assist Tomogr 8: 514–522

Kuhl DE, Phelps ME, Markham CH, Metter EJ, Riege WH, Winter J (1982) Cerebral metabolism and atrophy in Huntington's disease determined by ^{18}FDG and computed tomographic scan. Ann Neurol 12: 425–434

Kuwert T, Lange HW, Langen K-J, Herzog H, Aulich A, Feinendegen LE (1990) Cortical and subcortical glucose consumption measured by PET in patients with Huntington's disease. Brain 113: 1405–1423

Kuwert T, Hennerici M, Langen K-J, Aulich A, Herzog H, Sitzer M, Feinendegen LE (1991) Regional cerebral glucose consumption measured by positron emission tomography in patients with unilateral thalamic infarction. Cerebrovasc Dis 1: 327–336

Kuwert T, Sures T, Loken M, Langen K-J, Hennerici M, Feinendegen LE (1991) The influence of image resolution and of the size of regions of interest on the positron emission tomographic measurement of caudate glucose consumption. Nucl Med (submitted)

Laplane D, Levasseur M, Pillon B, Dubois B, Baulac M, Mazoyer B, Tran Dinh S, Sette G, Danze F, Baron JC (1989) Obsessive-compulsive and other behavioural changes with bilateral basal ganglia lesions: neuropsychological, magnetic resonance imaging and positron tomography study. Brain 112: 699–725

Levy AV, Laska E, Brodie JD, Volkow ND, Wolf AP (1991) The spectral signature method for the analysis of PET brain images. J Cereb Blood Flow Metab 11: A103–A113

Lueck C, Zeki S, Friston KJ, Delber M-P, Cope P, Cunningham VJ, Lammertsma AA, Kennard C, Frackowiak RSJ (1989) A colour centre in the cerebral cortex of man. Nature (London) 340: 386–389

Marrett S, Evans AC, Collins L, Peters TM (1989) A volume of interest (VOI) atlas for the analysis of neurophysiological image data. Proc SPIE 1092: 467–472

Martinot J-L, Hardy P, Feline A, Huret J-D, Mazoyer B, Attar-Levy D, Pappata S, Syrota A (1990) Left prefrontal glucose hypometabolism in the depressed state: a confirmation. Am J Psychiatry 147: 1313–1317

Mazziotta JC, Koslow SH (1987) Assessment of goals and obstacles in data acquisition and analysis from emission tomography: report of a series of international workshops. J Cereb Blood Flow Metab 7: S1–S31

Mazziotta JC, Phelps ME, Plummer D, Kuhl DE (1981) Quantitation in positron emission computed tomography. 5. Physical-anatomical effects. J Comput Assist Tomogr 5: 734–743

Mazziotta JC, Pelizzari CC, Chen GT, Bookstein FL, Valentino D (1991) Region of interest issues: the relationship between structure and function in the brain. J Cereb Blood Flow Metab 11: A51–A56

McNamara D, Horwitz B, Grady CL, Rapoport SI (1987) Topographical analysis of glucose metabolism, as measured with positron emission tomography, in dementia of the Alzheimer type: use of linear histograms. Int J Neurosci 36: 89–97

Mintun MA, Fox PT, Raichle ME (1989) A highly accurate method of localizing neuronal activity in the human brain with positron emission tomography. J Cereb Blood Flow Metab 9: 96–103

Moeller JR, Strother SC, Sidtis JJ, Rottenberg DA (1987) The scaled sub-profile model: a statistical approach to the analysis of functional patterns in positron emission tomographic data. J Cereb Blood Flow Metab 7: 649–658

Murayama H, Nohara N, Tanaka E, Hayashi T (1982) A quad BGO detector and its timing and positioning discrimination for positron computed tomography. Nucl Instr Meth 192: 501–511

Nutt R, Casey M, Carrol LR, Dahlborn M, Hoffman EJ (1985) A new multicrystal two dimensional detector block for PET. J Nucl Med 26: P28

Pelizarri CA, Chen GTY, Spelbring DR, Weichselbaum RR, Chen C-T (1989) Accurate three-dimensional registration of CT, PET and MR images of the brain. J Comput Assist Tomogr 13: 20–26

Rota Kops E, Herzog H, Schmid A, Winkens A, Dick R, Feinendegen LE (1989) Influence of some instrumental parameters on the determination of quantitative data using the PET scanner PC4096-15WB. Eur J Nucl Med 15: 701–704

Rota Kops E, Herzog H, Schmid A, Holte S, Feinendegen LE (1990) Performance characteristics of an eight-ring whole body PET scanner. J Comput Assist Tomogr 14: 437–445

Rottenberg DA, Moeller JR, Strother SC, Dhawan V, Sergi ML (1991) Effects of percent thresholding on the extraction of [^{18}F]fluoro-deoxyglucose positron emission tomographic region-of-interest data. J Cereb Blood Flow Metab 11: A83–A88

Seitz RJ, Bohm C, Greitz T, Roland PE, Ericksson L, Blomqvist G, Rosenqvist G, Nordell B (1990) Accuracy and precision of the computerized brain atlas program for localization and quantification in positron emission tomography. J Cereb Blood Flow Metab 10: 443–457

Strother SC, Liow J-S, Moeller JR, Sidtis JJ, Dhawan VJ, Rottenberg DA (1991) Absolute quantitation in neurological PET. Do we need it? J Cereb Blood Flow Metab 11: A3–A16

Tyler JL, Strother SC, Zatorre RJ, Alivisatos B, Worsley KJ, Diksic M, Yamamoto YL (1988) Stability of regional cerebral glucose metabolism in the normal brain measured by positron emission tomography. J Nucl Med 29: 631–642

Valentino DJ, Mazziotta JC, Huang HK (1988) Mapping brain function to brain anatomy. Proc SPIE 914: 445–451

Young AB, Penney JB, Starosta-Rubinstein S, Markel DS, Berent S, Giordani B, Ehrenkaufer R, Jewett D, Hichwa R (1986) PET scan investigations of Huntington's disease: cerebral metabolic correlates of neurological features and functional decline. Ann Neurol 20: 296–303

Authors' address: Dr. T. Kuwert, Institute of Medicine, Research Center Jülich, P.O. Box 1913, D-W-5170 Jülich, Federal Republic of Germany

J Neural Transm (1992) [Suppl] 37: 67–78

The role of anatomic information in quantifying functional neuroimaging data

C. Bohm[1], T. Greitz[2], and L. Thurfjell[3]

[1] Department of Physics, University of Stockholm, [2] Departments of Neuroradiology, Karolinska Institute/Hospital, Stockholm, and [3] Centre for Image Analysis, Uppsala University, Uppsala, Sweden

Summary. When using modern neuroimaging tools, such as CT, PET, SPECT, MRI and MEG, in brain research and brain diagnostics, there is a common need for including external anatomical information into the interpretation and analysis of data. This information may be used to aid the interpretation of structures in images from low resolution imaging tools. With high resolution tools it can help to identify resolved structures. It can also facilitate the merging of data from different modalities, or from different individuals. The anatomical information is often given as regions of interests (ROIs), which may be manually created from an anatomy rich image or automatically created from a standard template collection or from an atlas data base. Automatic methods will lead to a substantial reduction in bias and in size of the systematic errors. Functional ROIs can correspondingly be derived from functional images (usually PET or SPECT). Different aspects of these processes are discussed in the report.

Introduction

During the last decade several new diagnostic tools have been developed for examining the human brain. Many of these deliver point-by-point data corresponding to local values of different physical parameters. Such parameters or combinations of them can then be transformed into physiological or morphological data. The data are often organized in 2-dimensional data sets which may be visualized as images. These methods are thus often described by the common term "neuroimaging". Computed tomography (CT), magnetic resonance imaging (MRI) and positron emission tomography (PET) belong to this category. These tools have opened up new avenues within psychiatric research as they allow a qualitative interpretation of physiological or morphological conditions.

There are also reasons to believe that objective classifications of psychiatric disorders may be derived from statistical analysis of PET data. However, in order to statistically evaluate observations quantitative methods are required. Efforts to achieve accurate and reliable quantification in PET studies of human brain physiology have been made by different investigators (Bajcsy et al., 1983; Bohm et al., 1986, 1991; Evans et al., 1988; Fox et al., 1985, 1988; Friston et al., 1989, 1991; Greitz et al., 1989, 1991; Maziotta et al., 1981, 1984; Seitz et al., 1990). Many obstacles have yet to be overcome and a final solution of the problems involved is still awaiting.

Quantitation and regions of interests

The central task when quantitatively evaluating neuroimaging data is to extract a small number of representative values (statistics), which contain qualitative information relevant to the processes under investigation. If the statistics are efficient they extract information from all relevant picture elements (pixels). The values in these pixels, which usually form a closed structure, are summed or averaged. Such structures are often called regions of interest (ROI's).

The statistic should also be independent on variations among irrelevant parameters. Such variations occur when the data are obtained from different measurements, different neuroimaging equipments or different subjects. Even if the measurements are performed within a sufficiently short time interval so that no physiological or morphological changes have taken place, variations in patient positioning and uncorrected instrumental instabilities will influence the result. If different imaging equipments are used different scaling factors and effects due to different resolution functions are also introduced. If the study involves averaging and comparisons between different subjects, anatomical variations will be important (Fox et al., 1988).

Reducing the influence from irrelevant parameters

Some dependencies can be removed by also letting the ROI boundary depend on the parameter, but in such a way that the image and ROI variation induced by the parameter change counter-act. This is the case when the ROI is related to the center of the brain image and to the orientation of the interhemispheric fissure. A misalignment of the patients head will then move both data and ROI, leaving the statistics, i.e. the ROI average, unaffected. The remedy for eliminating some ROI average parameter dependancy is thus to demand that the ROI specification should be dependent on the parameter.

Another way to express the same process is to demand that the ROI should be specified in parameter independent coordinates. This may be realized by first transforming the data so that the parameter maps into a standard value before applying a predetermined ROI. For example, a ROI that is independent on the head position is achieved by first transforming the data so that the center of the brain image coincides with the center of the image itself (e.g. by expressing the ROI in center of brain coordinates).

Anatomical variations can, at least in principle, be treated in a similar fashion. This means relating the ROI to anatomical boundaries (Bohm et al., 1986; Greitz et al., 1989). Information about such boundaries is, however, not always present in the data set. In this case the boundaries may be derived from other sources of information or approximately determined by some indirect method. Resizing the images in x-, y- and z-direction to fit the brain outline to a standard brain-shape before applying the ROI is one way to obtain approximate anatomy invariance.

Selecting regions of interest

Let us define a region of interest as a connected subset which contains data of a certain type.

This definition suggests a ROI quality measure, *specificity*, describing the ROI selectivity, and by that a way in which the ROI may fail: it may not be specific enough. If it is specific, then it only includes the designated data, and when it is not specific enough it also includes other data than the designated type.

A ROI may also be judged according to how *sensitive* it is. A sensitive ROI will include all data of the designated type. Specificity and sensitivity are closely related to systematic and statistical errors. Low specificity introduces systematic errors while low sensitivity causes large statistical errors. Clearly one desires both specificity and sensitivity, but these two concepts

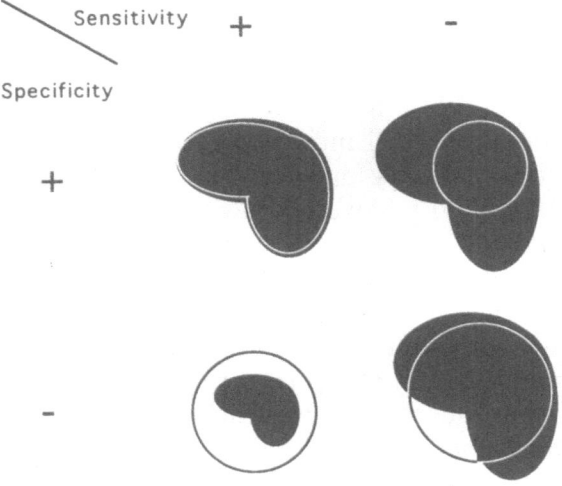

Fig. 1. ROI's with varying degrees of specificity and sensitivity

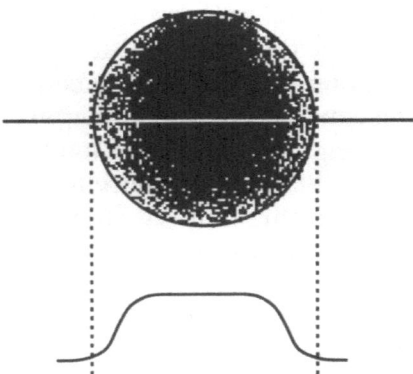

Fig. 2. If the ROI boundaries are not sharp, a sensitive ROI will not be specific

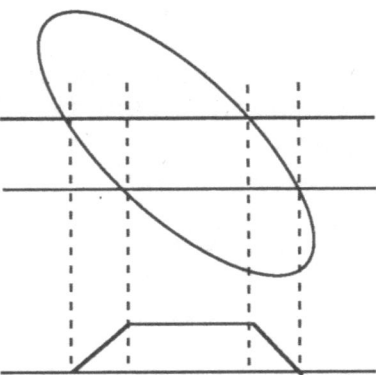

Fig. 3. If a 3-dimensional object is imaged by a method which has a finite and uniform z-resolution, each image will correspond to a slice. This will also result in boundary "fuzziness"

are often opposed to each other. A sensitive ROI is often not specific enough (and vice versa). A ROI may also be neither sensitive nor specific.

Trivial degrees of sensitivity or specificity can be achieved by expanding or contracting the ROI. A well-behaved ROI, however, will reach a reasonable degree of both sensitivity and specificity at the same time. Of all well-behaved ROI's with a certain degree of sensitivity the one with highest specificity is considered *optimal*. With sharp structure boundaries an optimal ROI is easily constructed by tracing the structure contour.

If the ROI boundary is not sharp (i.e. when the belonging function is a fuzzy set), a ROI may be formed by tracing an iso-contour of say 50% belonging. The level of belonging will affect the properties of the ROI. A high level will reduce the sensitivity and increase the specificity reducing the systematic errors at the expence of the statistical errors. A low level of belonging will cause the opposite result.

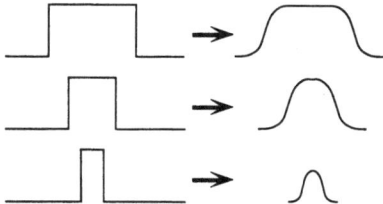

Fig. 4. Objects with sharp boundaries will be smoothed by the in-plane resolution of the measuring instrument

The systematic errors will be of two kinds: due to the overestimation of the amount of data of the correct type and due to the influence of data from surrounding structures. They will also depend on the operation performed on the ROI data. If the integral value of some parameter within the ROI is desired, the operation is summation. A sensitive ROI will in this case only cause systematic errors if information from surrounding tissue interfere. If, on the other hand, the average value within the structure is desired, the correct strategy under the same circumstances is to chose a specific ROI. Another way is to use a sensitive ROI and to correct for the underestimation caused by the exaggerated boundary. Such a correction factor is called a recovery coefficient (Mazziotta et al., 1981).

Structures with fuzzy boundaries are common in neuroimaging. They are introduced by the fact that the imaged objects are extended in the z-direction together with the finite z-resolution of the imaging apparatus. This is called the partial volume effect. Another source of fuzziness is the finite in-plane resolution of the apparatus.

If the imaged object is small compared to the instrument resolution, it is no longer possible to eliminate the systematic errors in averages by choosing a small ROI (see Fig. 4).

To derive a 2-dimensional ROI from a 3-dimensional structure specification, obtained from another high resolution morphological imaging method or from a brain atlas (see below), it is useful to calculate how the 3-d structure would be viewed by the instrument assuming the structure to be uniform (Bohm et al., 1991). The resulting distribution (which we may call a 2.5 dimensional ROI since it contains incomplete volume information) will describe how much the structure will contribute to the values in the different pixels in the imaging plane. The 2-d ROI is then derived from a tresholding operation on the 2.5-d ROI. The level should be set with regard to the experimental conditions.

A ROI, either 2- or 3-dimensional, may be defined by its anatomic boundaries or by its physiologic properties. The anatomically defined ROI is more often liable to fail by not being specific enough compared to the physiologic ROI, e.g. the entire motor area of the cortex cannot be used as ROI in an experiment designed to study the effect of stimulation of the hand area. On the other hand, a physiologic ROI is difficult to delimit prior

C. Bohm et al.

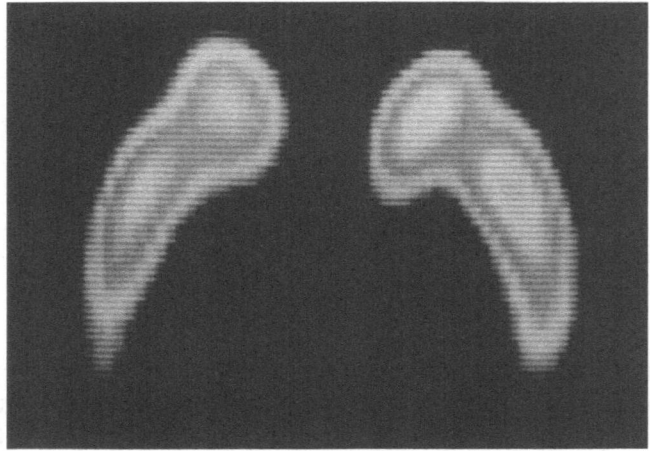

Fig. 5. A 2.5-dimensional ROI corresponding to the caudate nucleus and putamen viewed by a PET camera with the resolution of Scanditronix PC-384

to the experiment, especially when the aim is to map the various activity centers, and hence presupposes·the extent and location of the ROI to be unknown. Thus, any attempt to predefine the physiological ROI will fail and an empirical method of selecting such ROIs must be implemented (Friston et al., 1991).

ROI determination strategies

There exists several methods for ROI determinations with varying degrees of reliability.

Drawing anatomic ROI's in the image under study

This simple an straight-forward method has several disadvantages. It is inherently subjective and therefore not robust. With poor statistics the delineation will be affected by the noise introducing a bias. Furthermore, the fact that the regions are not 2-d but 3-d (partial volume effects) together with limited resolution will cause systematic errors and inconsistencies.

The disadvantages can, however, be reduced. The subjectivity may be reduced by making several independent evaluations and the bias is reduced when increasing the information content. The inconsistency introduced by partial volume effects may be corrected by introducing correction factors from calibrations, i.e. recovery coefficients. A strategy to reduce the partial volume effects at the expense of efficiency is to shrink the size of the ROI.

Combining several modalities

In the PET case, the bias may also be reduced by drawing the ROI's in corresponding slices from another modality such as CT or MR. The superior resolution of these modalities will increase both accuracy and precision. However, the method introduces new sources of error linked to the key-word "corresponding". Displacements or rotations of the CT or MR planes relative to the PET planes cause errors. These errors are reduced but not completely removed by using a fixation system to secure the subject.

Another way is to automatically identify the PET plane with a CT or MR plane or an interpolated plane by matching the skull outline (Maguire et al., 1991). This avoids cumbersome fixation systems but demands densely spaced CT or MR slices, i.e. large data sets (and quite demanding computer operations). Both methods are, however, relatively insensitive to pathological anatomies.

Atlas based methods

There are several methods based on computerized brain atlases, where 3-d maps are adapted (transformed) to the subject (Fig. 6). The ROI's are then derived from the map. Different methods use different maps and different transformations. The simplest approach is to estimate the rotation and scale to the correct size. A better approach is to fix the orientation from fiducial points (i.e. landmarks such as the AC-PC line) before the resizing. Even better is to use more general transformations (e.g. a general 2nd degree polynomial in all 3 dimensions) adapting many fiducial points (Bohm et al., 1983; Greitz et al., 1991). The best alternative is, however, to use a large number of fiducial points and to elastically deform the atlas between these points. The final strategy is not always applicable and is far from trivial. However, the problems will eventually be solved and used, independent of the computing power required. The future is in complex transformations, but an advantage with the simpler transformations is that they may be performed automatically, thus reducing the subjectivity.

In general one can say that the required atlas precision depends on the precision on the neuroimaging system. Thus, a low resolution method like PET is more permissive concerning errors in the atlas and corresponding transformation. However, atlas methods generally have difficulties with pathological anatomies. Computerized atlases also differ in that some contain detailed anatomical structures, while other are satisfied with simplified structures-templates.

Complex atlases may be adapted to low resolution data such as PET in two steps: the detailed transformation is first derived from an anatomy-rich modality, determining only the final rigid transformation from the PET image itself.

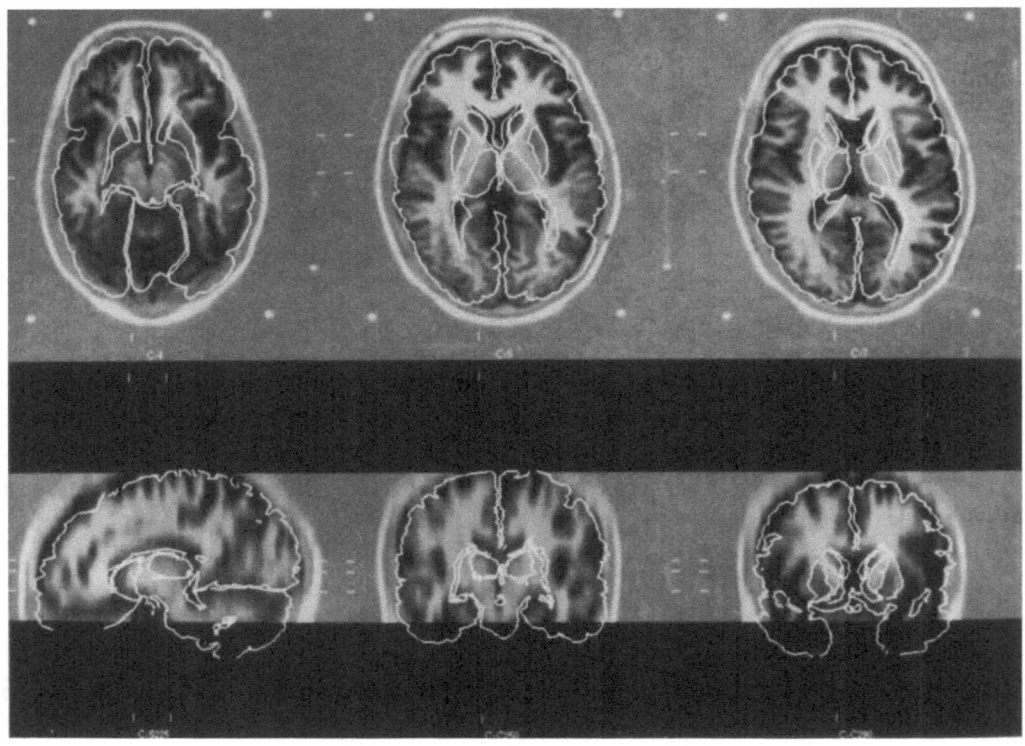

a

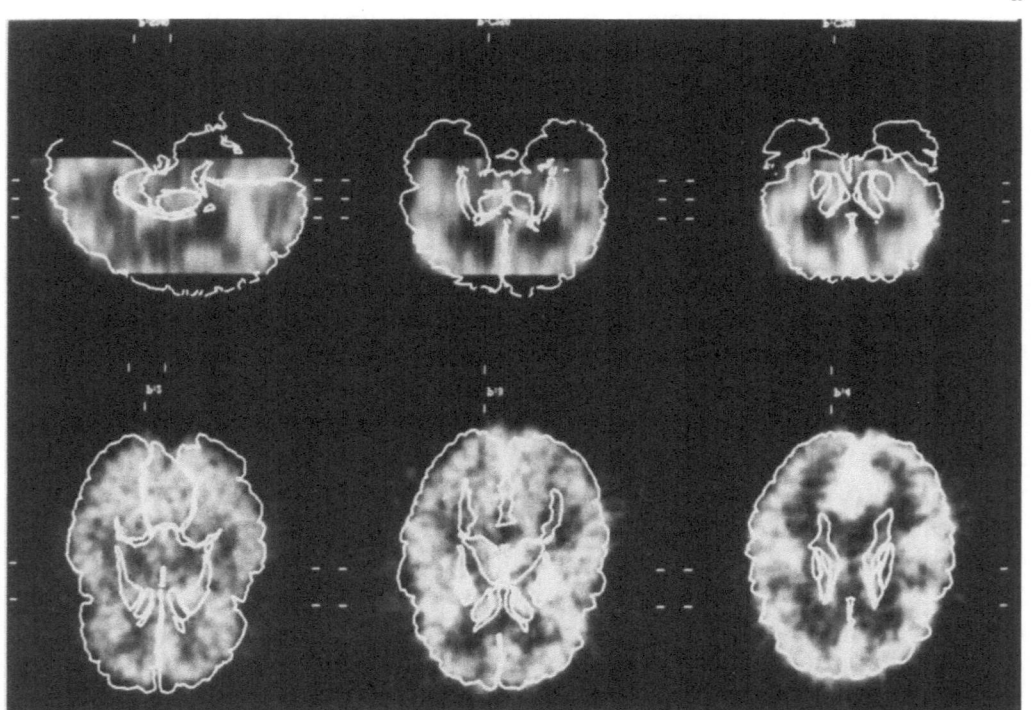

b

A useful application of an anatomic atlas is the possibility to simulate different measuring situations to investigate how the result is affected by different assumptions. An example of this is the determinations of recovery coefficients (Bohm et al., 1985, 1986, 1991). Such simulations may also provide a sensitivity analysis. Other uses of an atlas are linked to the possibility to transform structures and data sets to a reference system where they may be compared.

Comparing data from different subjects

One very frequent demand is to combine neuroimaging data from different subjects within a group into group averages (Bohm et al., 1986; Fox et al., 1988; Creitz et al., 1989). These averages may then be used to find significant differences between groups or to find how a given individual differs from different normal groups.

In order to form meaningful group averages, it is, as has already been pointed out, necessary to eliminate or to reduce the influence of anatomical variations. The two approaches described above lead to two methods: to merge individual anatomy-independent ROI averages or to calculate ROI averages from a merged data set where anatomy standardized data from different subjects are combined. The methods can also be described as averaging ROIs or ROIing averages.

The anatomy independent ROI's are obtained from the image itself, from an equivalent morphological image made in another imaging modality, or from an individualized (see below) computerized brain atlas. Image data are standardized by applying transforms to it that converts the anatomy of the subject into a standard (atlas) anatomy.

Other issues that must be resolved concern normalizing the individual data before the merge and compensating the pixel values when deforming the structures. Normalization may follow two extreme strategies. One is to neglect it, assuming that the local value is not affected by global considerations. The other approach is to normalize the local values so that the global value assumes a standardized value (Fox et al., 1988). The rationale for this is to assume that the global value is fixed but varying from subject to subject. An intermediate path is to estimate the proper normalization from the data using covariance analysis (Friston et al., 1989).

The transformation induced pixel value compensation depend on what ROI operation is performed. If the total sum is the desired statistic, a transformation causing the structure to expand must be compensated for by

---

Fig. 6. a The computerized atlas (Bohm et al., 1983, 1984; Greitz et al., 1991) has been adapted to a set of MR-images. **b** After reformation, the modified atlas was transferred to the patient's PET images. The adapted structures include the brain surface, the ventricular system, the striatum and the thalamus

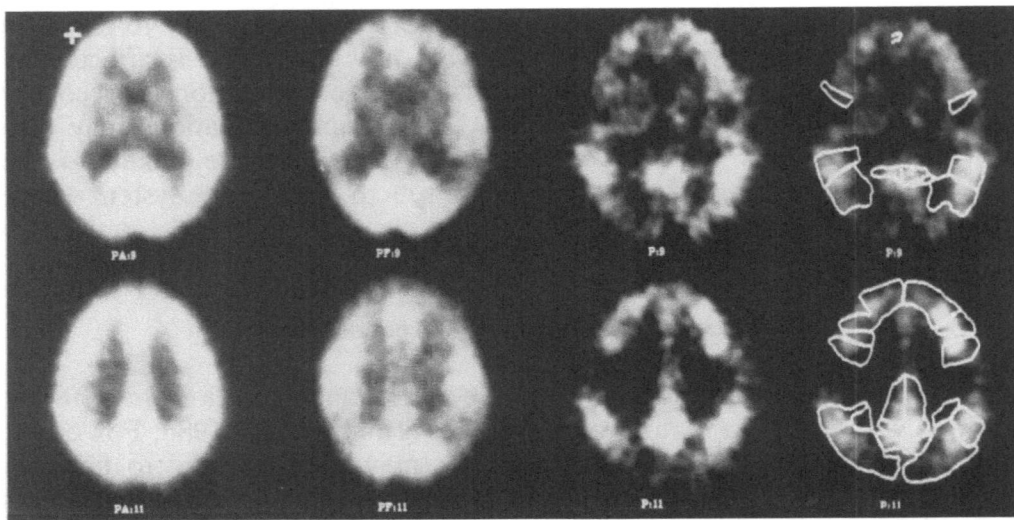

Fig. 7. The cerebral metabolic rate of glucose as measured with PET is shown in a clinical case of dementia, using the "individual-mean subtraction method" according to Greitz et al. (1991). Brain slices at two levels are shown (upper and lower row). The two images to the left represent mean images obtained after standardization of each individual anatomy in a group of nine age-matched controls and averaging of the standardized images. The other image-pairs are from left to right: reformatted images of the patient, subtraction images (mean minus patient) without and with (to the right) the most affected Brodmann areas drawn in by the atlas program. These are areas 6, 7, 8, 9, 19, 23, 34, 39, all bilaterally

a reduction of the value (mathematically by including the jacobian of the transformation) in order to maintain the overall value. If the average is the intended operation no compensation is necessary.

Some atlas based methods allow for the image data to be transformed into a standardized anatomy. The data from different subjects can thus be compared on a pixel-by-pixel basis rather than region-by-region as in the first method refered to as "averaging ROIs". The result of these operations can be visualized. Images showing group averages can be created, and differences between various group averages or between one individual and a group average (Fig. 7) can be visualized in subtraction images.

Functional regions of interest

Functional imaging devices, such as PET, combined with the possibility to reformat data into a standardized anatomy by the use of a computerized brain atlas, are excellent tools in the process of mapping functional areas of the brain.

The simplest approach in the study of physiological stimulation is to use paired-image subtraction (task minus control). In this type of experiment two scans are obtained for each subject, one during rest and one during physiological stimulation.

We may write a statistical model that describes the data from the experiment as

$$y_{ij} = \mu_i + \beta_j + \varepsilon_{ij} \qquad (1)$$

y_{ij} is the observation (i.e. the pixel value) obtained during scan i on subject j, μ_i is the true response, β_j is an effect due to subject j and ε_{ij} is an experimental error (Montgomery, 1983). By subtracting the image at rest from the image obtained during stimulation we get

$$d_j = y_{1j} - y_{2j} \qquad (2)$$

The expected value of this difference is

$$m_d = E(d_j) = E(y_{1j} - y_{2j}) = E(y_1 d_{2j}) - E(y_{2j}) = (m_1 + \beta_j) - (m_2 + \beta_j)$$
$$= \mu_1 - \mu_2$$

and the null hypothesis that no change has occurred at the specific location is $H_0 : \mu_d = 0$. Note that the additive effect of the individuals in (1) cancels out in (2). The test statistic for this hypothesis is then computed as

$$t = \frac{d}{S_d / \sqrt{n}}$$

where d is the sample mean differences and S_d is the sample standard deviation of the differences.

Note that the procedure above only describes the computation of the t statistic in one location (i.e. for one pixel), and it is repeated for all pixels in the image thus creating a statistical parametric map (SPM) of the t statistic. However, even though this simple model is excellent for visualizing and identifying areas with increased brain activity, it is not possible to use the results for hypothesis testing (i.e. finding those pixels that shows an increase in brain activity at a certain level of significance) and the method is thus not applicable for outlining functional areas of the brain. This is mainly due to the fact that the large number of comparisons made are not independent. One source for dependence between adjacent pixels is, as previously mentioned, the finite in-plane resolution of the apparatus. One attempt to overcome this obstacle by using an estimation of the smoothness in the SPMs to calculate the threshold required to identify significant foci has recently been reported by Friston et al. (1991).

Conclusion

Although many problems have been overcome, a standardized technique for analyzing data from physiological experiments has yet to be established before the ultimate goal of creating a three-dimensional functional brain atlas can be achieved.

References

Bajcsy R, Lieberson R, Reivich M (1983) A computerized system for the elastic matching of deformed radiographic images to idealized atlas images. J Comput Assist Tomogr 7: 618–625

Bohm C, Greitz T, Kingsley D, Berggren BM, Olsson L (1983) Adjustable computerized stereotaxic brain atlas for transmission and emission tomography. Am J Neuroradiol 4: 731–733

Bohm C, Greitz T, Berggren BM (1985) Selection of PET ROIs from a computerized brain atlas. J Cereb Blood Flow Metab 5 [Suppl] 1: S613–S614

Bohm C, Greitz T, Blomqvist G, Farde L, Forsgren PO, Kingsley D, Sjögren I, Wiesel FA, Wik G (1986) Applications of a computerized adjustable brain atlas in positron emission tomography. Acta Radiol [Suppl] 369: 449–452

Bohm C, Greitz T, Seitz R, Eriksson L (1991) Specification and selection of regions of interest (ROIs) in a computerized brain atlas. J Cereb Blood Flow Metab 11: A64–A68

Evans AC, Beil C, Marett S, Thompson C, Hakim A (1988) Anatomical-functional correlation using an adjustable MRI-based region of interest atlas with positron emission tomography. J Cereb Blood Flow Metab 8: 513–530

Fox PT, Perlmutter JS, Raichle ME (1985) A stereotactic method of localization for positron emission tomography. J Comput Assist Tomogr 9: 141–153

Fox PT, Mintun MA, Reinman EM, Raichle ME (1988) Enhanced detection of focal brain responses using intersubject PET images. J Cereb Blood Flow Metab 8: 642–653

Friston KJ, Frith CD, Liddle PF, Lammertsma AA, Dolan RD, Frackowiak RSJ (1989) The relationship between local and global changes in PET scan. J Cereb Blood Flow Metab 10: 458–466

Friston KJ, Frith CD, Liddle PF, Frackowiak RSJ (1991) Comparing functional (PET) images: the assessment of significant change. J Cereb Blood Flow Metab 11: 690–699

Greitz T, Bohm C, Eriksson L, Mogard J, Roland PE, Seitz RJ, Wiesel FA (1989) The construction of a functional brain atlas: elimination of bias from anatomical variations at PET by reformatting three-dimensional data into a standardized anatomy. In: Ottoson D, Rostene W (eds) Visualization of brain functions. Macmillan, London, pp 137–140 (Wenner-Gren Center International Symposium, vol 53)

Greitz T, Bohm C, Holte S, Eriksson L (1991) A computerized brain atlas: construction, anatomical content, and some applications. J Comput Assist Tomogr 15: 26–38

Maguire GQ, Noz ME, Rusinek H, Jaeger J, Kramer EL, Sanger JJ, Smith G (1991) Graphics applied to medical image registration. Comput Graph Appl 2/2: 20–28

Mazziotta JC (1984) Physiologic neuroanatomy. New brain imaging methods present a challenge to an old discipline. J Cereb Blood Flow Metab 4: 481–483

Mazziotta JC, Phelps ME, Plummer D, Kuhl DE (1981) Quantitation in positron emission computed tomography. 5. Physical-anatomical effects. J Comput Assist Tomogr 5: 734–743

Montgomery DC (1983) Design and analysis of experiments, 2nd edn. Wiley, New York, pp 32–36

Seitz RJ, Bohm C, Greitz T, Roland PE, Eriksson L, Blomqvist G, Rosenkvist G, Nordell B (1990) Accuracy and precision of the computerized brain atlas program for localization and quantification in positron emission tomography. J Cereb Blood Flow Metab 10: 443–457

Authors' address: Dr. C. Bohm, Department of Physics, University of Stockholm, Vanadisvägen 9, S-113 46 Stockholm, Sweden

J Neural Transm (1992) [Suppl] 37: 79–93
© Springer-Verlag 1992

The dorsolateral prefrontal cortex, schizophrenia and PET

K. J. Friston

(for the Neuropsychiatric Group*, Clinical Sciences Section, MRC Cyclotron Unit and Royal Postgraduate Medical School)

MRC Cyclotron Unit, Hammersmith Hospital and University Department of Psychiatry, Charing Cross and Westminster Medical School, London, United Kingdom

Summary. Central neurophysiology can be measured with PET. These measurements are providing insights into the regional abnormalities associated with schizophrenia. Cohorts of schizophrenic subjects have been studied cross-sectionally in attempts to identify common regional deficits. More recently the advent of fast dynamic measurements of regional cerebral blood flow have allowed rapid serial measurements in the same subject in different brain states (activation studies). These complementary approaches are based upon, and are interpreted with reference to, a number of methodological considerations and underlying hypotheses. The key hypotheses underpining cross-sectional and activation studies are discussed within the framework of the lesion model and functional anatomy models of brain function. This brief review of some assumptions, ideas and methodological constraints is illustrated with empirical data implicating the dorsolateral prefrontal cortex in schizophrenic symptoms.

Introduction

Cognitive activation studies of neurophysiology and cross-sectional studies of psychiatric cohorts are two approaches to the study of brain metabolism and blood flow in psychiatric disorder with positron tomography (PET). These approaches are predicated on different models relating neurophysiology and behaviour, namely functional anatomy and the disease [or lesion] model. The interdependency of these approaches is discussed. This discussion is illustrated with studies of schizophrenic and normal subjects which focus on the left dorsolateral prefrontal cortex.

---

*C. J. Bench, R. D. Dolan, R. S. J. Frackowiak, C. D. Frith, K. J. Friston, P. Grasby, P. F. Liddle

Methodology

PET activation studies rely on rapid, serial estimation of regional cerebral blood flow (rCBF) using a dynamic $C^{15}O_2$ technique. This measurement typically takes a minute or so (Lammertsma et al., 1990). The images of rCBF resolve the brain into 8 mm elements (Spinks et al., 1988). It is gernerally accepted that rCBF reflects regional neural activity.

The objective of image analysis is to identify brain areas that evidence a change in activity between behavioural states or between different diagnostic groups. The data presented in this discussion were analyzed using statistical parametric mapping (SPM). SPM refers to the construction of parametric maps whose voxel [or volume element] values are distributed according to a statistic. The most commonly used is the t statistic to compare two condition [group] means. Other examples of SPMs include the correlation coefficient and the F ratio. In the analysis of activation studies pairwise (and non-pairwise) comparisons between combinations of different brain states are done using a test quotient with the t distribution. In comparing two cohorts the t statistic can be used, alternatively correlational analyzes can be performed using parametric maps of the correlation coefficient. The stages of analysis [in this unit] which follow acquisition of rCBF images include:

1. Attenuation correction: A correction is made for the attenuating effects of the skull and intracranial tissue on "emitted" photons using measurements of the attenuation of "transmitted" photons from an external source.

2. Stereotactic normalization: The brain image is repositioned, resized and reshaped into a standard stereotactic space (Talaraich and Tournoux, 1988). Repositioning (translation and rotation) proceeds with reference to a standard anatomical plane which passes through the anterior and posterior commissures (AC-PC plane). The AC-PC plane is estimated directly from morphological information in the primary (PET) image. This approach has been validated by comparison to techniques which use cross-modal registration of functional (PET) and structural (skull X-ray) images (Fox et al., 1987; Friston et al., 1989).

Linear resizing uses the edges of the brain and assumes a proportional stereotactic space. A final non-linear or plastic resampling of the image accounts for differences in gross brain shape (e.g. asymmetry) that remain after linear resizing. This plastic resampling does not use landmarks but distorts the space of the observed image to maximize correspondence with a standard template for the slice in question (Friston et al., 1991a). The resampling is effected by replication and therefore preserves point concentration of activity but not global indices. For example, if two subjects have the same cortical rCBF but one subject's brain was 10% bigger, following stereotactic normalization both subjects would have the same regional and global CBF.

3. Global normalization: The confounding effects of global (whole brain) differences on regional values are covaried out using analysis of covariance for each voxel. This assumes that global differences do not markedly impact on the activation effect or the regional difference between two groups (the global effect is additive, not multiplicative). This regression analysis generates mean activities for each activation condition [group] and an associated error variance for each voxel. Clearly an additive model of the relationship between regional and global values is an approximation to a non-linear relationship, but probably the best approximation (Friston et al., 1990).

4. Comparison of image means: [non-]pairwise comparisons of the mean activities for each condition [group] uses a test quotient with the t distribution. This step creates a map of t values (SPM{t}). This is an image of change significance. About 5,104 voxels constitute the data set. There are no further data transformations. By using an estimate of error variance for each voxel, this approach allows for heterogeneity of error variance [consistency of the difference] over the brain.

5. Assessment of significance: This relies on rejecting the null hypothesis that a distribution of voxel values could have arisen by chance. In the case of SPM{t} the distribution under the null hypothesis is known (t). The number of voxels above an arbitrary threshold (usually $p = 0.001$) is compared with that expected. This subset [excursion set] of suprathreshold voxels can constitute a significant profile in an omnibus sense if the observed number of voxels in the excursion set exceeds chance expectation. Omnibus in this context refers to the fact a single null hypothesis, relating to a collection of voxels, is being tested. This collection of voxels can also represent a series of non-independent multiple univariate tests (see Friston et al., 1991b, for a fuller discussion).

The SPM is displayed as a profile by presenting the highest t value along any line of view in three orthogonal projections of the brain space analyzed. This is effectively an X-ray of statistically dense (significant) regions. The significant subset of profiles is defined by the arbitrary threshold and only voxels exceeding this threshold are shown. The profiles are interpreted neuroanatomically by referring to the atlas (Talaraich and Tournoux, 1988).

Schizophrenia and PET

Two ways of using PET in schizophrenia research are to identify "relevant" functional anatomy in normals and to search for specific regional deficits in patients. These approaches use longitudinal activation studies and cross-sectional single state studies, respectively. Both are designed to find the behavioural correlates of regional neurophysiology. Activation studies introduce differences in neurophysiology through manipulations of behaviour. Cross-sectional studies introduce differences by comparing cohorts which do and do not show behavioural [clinical or neuropsychological] abnormalities.

The relationship between stimulation experiments and disease or lesion studies has a long history in neuroscience. A landmark meeting, that took place on the morning of August 4th, 1881 to discuss localization of function in the cortex cerebri, addressed this issue. Goltz (1881), although accepting the results of electrical stimulation of the dog and monkey cortex (e.g. Ferrier, 1875), considered the excitation method inconclusive, in that movements elicited might have originated in related pathways or current could have spread to distant centres. "Ablation experiments were therefore essential to complement the results obtained by excitation" (Phillips et al., 1984). The importance of the relationship between the behavioural correlates of cortical lesions and stimulation of viable cortex persists to the present day. This relationship is discussed in terms of PET, the disease model and functional anatomy.

The disease model

Despite advances in neuroscience over the past century, the most suggestive evidence for functional localization still derives from the sort of observations that impressed scientists a hundred years ago. For example the report by Zihl et al. (1983) of a patient with a bilateral prestriate lesion demonstrated by CT scan who was unable to detect motion but was able to read, see colour, form and depth, was a most convincing account in support of functional localization (Phillips et al., 1984). PET has now provided the complementary excitatory evidence (Zeki et al., 1991).

In the disease model behavioural abnormalities are associated with a specific cerebral dysfunction. In the case of schizophrenia an unknown pathophysiology can be suggested as a part cause of clinical and neuropsychological symptoms secondary to selective impairment of cortical and subcortical structures. This approach has its limitations. Indeed, early ablation studies were confounded by observed restitution of function, which can now be understood in terms of neuroplasticity and recruitment of unaffected areas. It is possible that the behavioural manifestation of a cerebral lesion, be it ultrastructural, chemical, atrophic etc., reflects loss of function ascribed to the region lost, or an interaction between loss and compensatory changes. A second problem with the lesion model relates to the massive connectivity and distributive nature of neural organization. Although brain systems may demonstrate functional specificity (Lueck et al., 1989) it is probable that many "functions" are topographically distributed (Mesulam, 1990).

In the disease model two levels are measurable: neurophysiology [e.g. rCBF with PET] and behaviour [with neuropsychological tests and clinical rating scales]. It is not unreasonable (given the above qualifications) to expect that these measures should be correlated in, and only in, affected brain systems underlying the behavioural abnormality.

The structure of schizophrenia

One confounding factor in schizophrenia research is heterogeneity. If schizophrenia was a unidimensional disorder [characterized by a single score], it would be a relatively simple matter to find the brain region whose neurophysiological abnormality best correlated with severity. The functional anatomy of this system could then be inferred from the characteristic neuropsychological deficits of [a unidimensional] schizophrenia and confirmed using activation studies in normal subjects. However, schizophrenia is probably multidimensional [factor analytic studies (Bilder et al., 1985; Liddle, 1987; Mortimer et al., 1990; Arndt et al., 1991) would suggest three-dimensional]. This suggests that a number of brain systems are affected and, because they are distributed (Mesulam, 1990; Goldman-Rakic, 1988), they may overlap topographically. This leads to the possibility that a regional neurophysiological index could have significant partial correlations with several dimensions of schizophrenia, but because severity on those dimensions are uncorrelated these regions are never identified. One approach to this problem is to correlate measures of neurophysiology with each behavioural dimension separately by treating the subdimensions as separable but related disorders. Our work (Liddle et al., 1990) uses this approach.

Study 1 — Cross-sectional study of schizophrenia

We have studied 30 DSM-III-R (American Psychiatric Association, 1986) chronic schizophrenic patients all under the age of 55. The selection criteria placed an emphasis on persistent and stable symptoms. Symptom ratings were made using CASH (Andreasen, 1987) and then subject to factor analysis. This analysis confirmed a three dimensional structure to the behavioural data: psychomotor poverty, characterized by poverty of speech, movement and feeling; a disorganization syndrome coloured by inappropriate affect and incoherent speech with little informational content; and finally a dimension of positive experiential symptoms including delusions and hallucinations. Using the factor scores from this analysis as estimates of severity it was possible to separately correlate these three behavioural scores with rCBF. In this study the [three] SPMs created were parametric maps of the correlation coefficient between adjusted [for global differences] activity and behavioural score at every voxel (mathematically, this is equivalent to the partial correlation between rCBF and behavioural score having partialled out the effect of global or whole brain differences).

Using this approach we have demonstrated "hypofrontality" (reduced rCBF in the dorsolateral prefrontal cortex — DLPFC) in patients who showed poverty of speech, movement and affect [psychomotor poverty] but not of patients who suffer experiential symptoms (delusions and hallucinations) (see Fig. 1a).

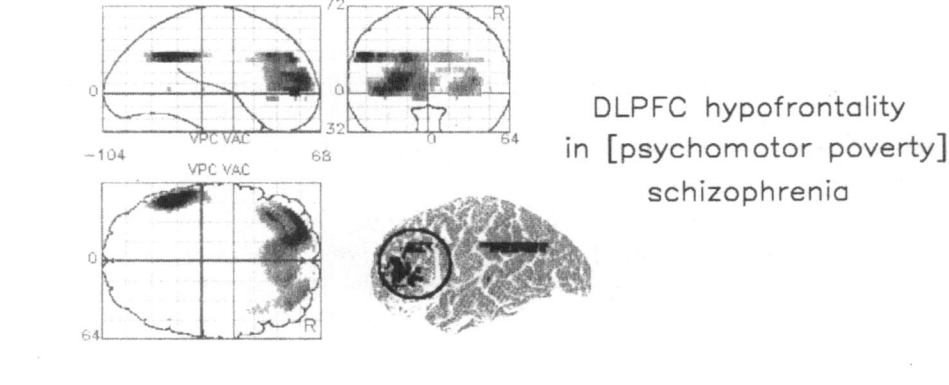

DLPFC hypofrontality
in [psychomotor poverty]
schizophrenia

a

DLPFC activation
during verbal fluency

DLPFC hypofrontality
following buspirone

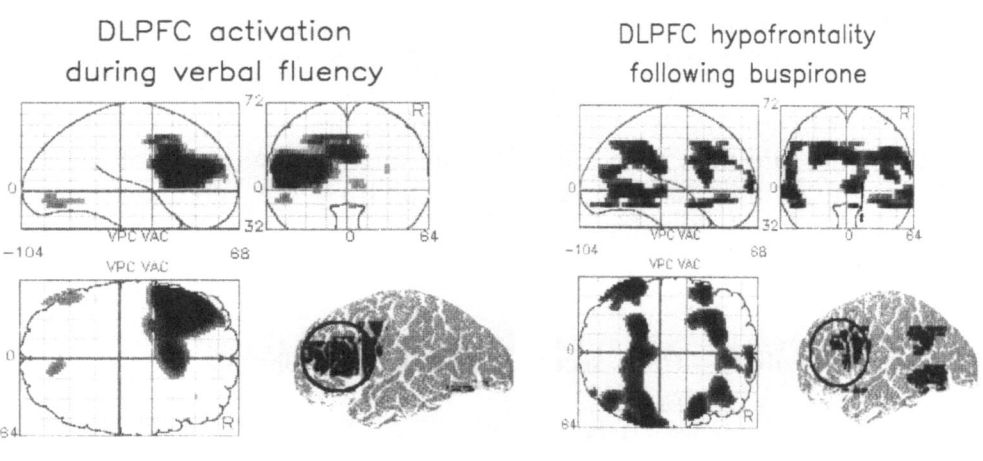

b   c

Fig. 1. Three statistical parametric maps (SPMs) displayed as volume images. The three boxes in each SPM represent views of the brain from the back, the right hand side and from the top. This proportional stereotactic space (Fox et al., 1985; Friston et al., 1989) corresponds to that described in the atlas of Talaraich and Tournoux (1988). The brightest pixel value along any line of view is displayed. The pixel value is a statistic which reflects significance. The same data has been rendered onto a drawing of the left lateral surface of the cortex. The DLPFC is circled for clarity. **a** SPM of significant (p < 0.05) negative correlations between rCBF and a "psychomotor poverty" score derived from a factor analysis of symptom scores. This psychomotor poverty score loads heavily on poverty of speech. The data are from thirty DSM-III-R chronic schizophrenic patients studied with the ^{15}O steady-state technique (Frackowiak et al., 1980). **b** SPM of significant (p < 0.001) t test values which test the difference in average rCBF on comparing a paced verbal fluency tasks with simply repeating a target word. The data are from 6 normal male subjects studied with a fast dynamic ^{15}O technique (Lammertsma et al., 1990). **c** SPM of significant (p < 0.001) t test values which test the significance of reduction in rCBF before and after the oral administration of 30 mg of buspirone, a partial 5-HT$_{1A}$ receptor agonist. The data are from 6 normal male subjects studied with a fast dynamic ^{15}O technique (Lammertsma et al., 1990)

Hypofrontality

Hypofrontality (Ingvar and Franzén, 1974) has been one of the more robust findings in functional imaging studies of schizophrenia. However there have been divergent findings. DeLisi et al. (1985a) compared chronic schizophrenic and control subjects. The patients were free of medication for at least two weeks before the study. Patients had significantly lower anterior-posterior gradients (hypofrontality) than the control group. Cerebral atrophy, as determined by CT was not correlated with this aberrant pattern. Gur et al. (1987a,b) has described abnormalities of cortico-subcortical metabolic gradients in schizophrenic patients but found no evidence for hypofrontality. Finally Szechtman et al. (1988) examined whether the duration of treatment influenced the regional distribution of metabolism in patients with schizophrenia. The schizophrenic group was dichotomized according to treatment duration. Both groups evidenced hyperfrontality when compared to controls. This was less evident in the group with the longest exposure to medication. One explanation for these inconsistencies may be the relative amounts of each behavioural subdimension the patient groups expressed. One can find a partial resolution in our findings, in that both normal frontality or hyperfrontality (associated with experiential symptoms) and hypofrontality (associated with psychomotor poverty) can co-exist in the same patient. This coexistence being apparent at both a behavioural and a neurophysiological level. Interestingly, DeLisi et al. (1985b) reported that the only significant correlations between relative hypofrontality and symptom ratings were for emotional withdrawal, disorientation, distractibility and helplessness/hopelessness.

The DLPFC does not appear to contribute to attentional aspects of behaviour, indeed lesions of the DLPFC in non-human primates generally improve performance on sensory discrimination tasks (Irle, 1990). The DLPFC has been specifically associated with response selection in the absence of extrinsic information (Goldman-Rakic, 1986). This is a definition of intrinsically generated, volitional, willed or intentional behaviour. Psychomotor poverty is characterized by intentional deficits, in the generation of intentional set and the translation of this into motor behaviour.

To complete a three-way link between abnormal DLPFC neurophysiology, psychomotor poverty and the functional anatomy of the DLPFC, it was necessary to show the DLPFC is critically involved in intentional behaviours that are impaired in psychomotor poverty (e.g. spontaneous speech and movement). This was done using activation studies in normal subjects.

The functional anatomy model

The functional specialization of brain systems can be established using activation PET studies in normal subjects. Strong predictions can be made about regional deficits in schizophrenia based on the characteristic neuropsychological deficits and functional anatomy (what is done where in the brain) defined in normal subjects. Alternatively strong predictions can be made about functional specialization given the regional neurophysiological deficit of a group that have a specific cognitive failure. Performing standard neuropsychological tests during rCBF measurements has proved an extremely fruitful strategy (e.g. Weinberger et al., 1986). However, there is a trend away from using standard tests and towards more carefully controlled tasks in which specific cognitive components can be dissected out. Tasks are designed following the idea of "cognitive subtraction" (Petersen et al., 1988; Posner et al., 1989). This idea has proved invaluable in the design and interpretation of cognitive activation studies using PET. It does however have limitations. The logic of cognitive subtraction relies heavily on the validity of pure insertion of a cognitive component. The demonstration of pure insertion using neuropsychological and psychophysical experiments has not been easy. Adding [or subtracting] a cognitive process may cause a change in strategy which changes pre-existing components rendering the cognitive difference a complicated one. From the point of view of schizophrenia research there is an added consideration, namely it is difficult to model some schizophrenic symptoms in normal subjects. Neuropsychological processes are measurable in schizophrenia and can be engaged in normal subjects, however schizophrenia manifests as clinical symptoms, some of which do not have obvious neuropsychological correlates [e.g. anhedonia, inappropriate affect, hallucinations and disorganized thought]. Intentional deficits are a key element of psychomotor poverty in schizophrenia and reflect processes that can be modelled in normal subjects. Intentional behaviour has been the focus of a series of experiments in this unit.

Intentional processes

Intentional behaviour was defined above as behaviour that is not extrinsically specified at the time of responding. By definition, intentional behaviour [in this sense] is contingent on past experience and is inherently mnemonic in character. An alternative definition of motor intention relates to set. Wise (1989) discusses the relationship between intention, attention and set. In one formulation set is seen as supreme and segregates into motor set and perceptual set, corresponding to motor intention and selective attention, respectively. In another formulation attention is considered hierarchically superior and can be subdivided in motor and sensory aspects. In both formulations, motor intention represents a state of preparedness to act which predates the act itself, again emphasising the central feature of behavioural organization across time. Work in non-human primates has implicated the prefrontal cortex in acquisition of motor set. Parts of the frontal cortex (supplementary motor area — SMA) are required for self-paced movements in the absence of extrinsic cues (Passingham et al., 1989). The relationship between motor (intentional) set and frontal cortex has been explored in terms of delay period activity (of neuronal firing during the withholding of a motor response). For example, Wise notes "Delay period activity is widely distributed in the frontal cortex as well as in structures that provide its inputs and receive its outputs" (Wise, 1989).

Verbal fluency is a particularly good example of intentional behaviour [word finding]. Schizophrenics, especially those with negative features, are bad at verbal fluency (Allen and Frith, 1983). Our PET studies of verbal fluency (Friston et al., 1991c; Frith et al., 1991) have demonstrated the central role the DLPFC has in this, and other, intentional behaviours.

Study 2 — The functional anatomy of intrinsically cued behaviour

Two companion studies were conducted, each in 6 normal right-handed males. Each study comprised 6 scans with 3 conditions repeated in balanced order. The two studies addressed word production and finger movements, respectively. The three conditions included an extrinsically cued condition (baseline), a novel extrinsically cued condition, and an intrinsically cued condition. For the lexical study this involved: repeating aurally presented concrete, high frequency, words; producing the opposite of a heard concrete word; and a paced verbal fluency condition. The motor tasks were lifting one of two fingers. In the first condition the touched finger was moved, in the second the finger not touched was moved, and in the third condition either finger was moved at random (in this instance the touching was a time cue for when to move as opposed to what to move). The most remarkable result from these studies was a DLPFC activation in, and only in, the intrinsically cued conditions. In the lexical study this was left sided, in the motor study it was bilateral. The SPM{t} from the lexical study is seen in Fig. 1b. The commonality in these activation profiles suggests the effect is due to that which is common to both studies, namely the intentional nature of the task, not the task modality. The novelty of the task and attentional aspects cannot be invoked as an explanation given an absence of activation in the opposites conditions.

This [DLPFC] region activates, in normals, during a cognitive process that is impaired in schizophrenic patients with negative features. Furthermore patients with negative features evidence hypofrontality in the same cortical region.

In conclusion, both the disease and functional anatomy models provide complementary and consistent information and each looks to the other for confirmation. This is a strong argument in favour of using both, preferably with the same methodology [data acquisition, image analysis and stereotactic space] (see Fig. 2 for a schematic summary of this argument).

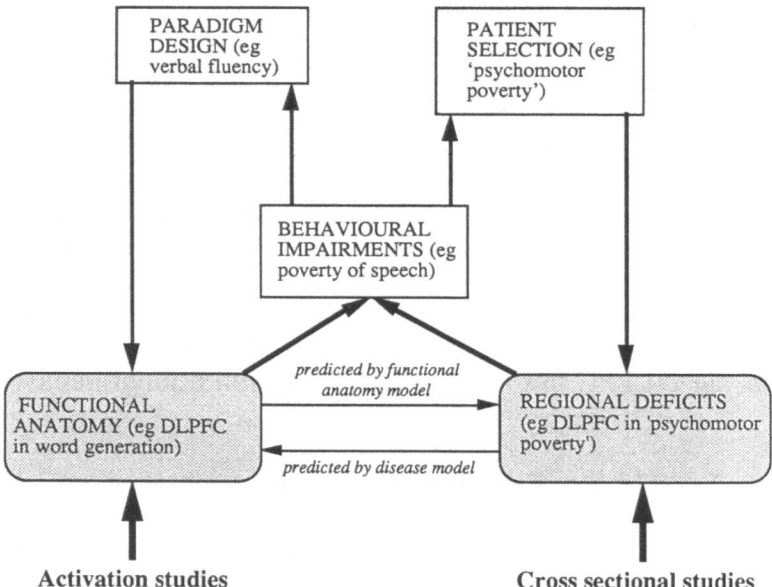

Fig. 2. Schematic illustrating the interdependency of cross sectional studies of psychiatric patients and the complementary activation studies of "relevant" functional anatomy in normal subjects

Functional anatomy in disease

It is a natural step to consider both models together, i.e. in activation studies of schizophrenic patients. The key advantage is that functional specificity is attributed directly to the regional deficit. The philosophy of demonstrating impaired response to challenge over and above baseline differences has been the tennent of neuroendocrine research for many years (e.g. Cowen et al., 1986).

There is, however, a difficulty when using tasks in subjects whose performance is likely to be impaired. This difficulty centers on the argument that a failure to activate cannot be used as an inference of regional dysfunction because a more parsimonious explanation is simply the subject did not perform. In our opinion this argument is vacuous but a more detailed analysis does raise some interesting points. In a simple sense, the impairment of self-paced movements in non-human primates following ablation of the SMA cannot be invoked as an explanation for dysfunction in the SMA. The question "does [motor] behaviour cause changes in central physiology or do changes in central physiology affect behaviour?" is simply answered — behaviour is secondary to central changes. A failure to activate a brain region in association with impaired performance cannot be explained by the assertion "this region did not activate because there was impaired motor behaviour". However it could be suggested that the stimulus to this area was in some way inadequate. This argument highlights the special case cognitive activation studies represent. Unlike electrical stimulation in animals or pharmacological stimulation in man, the efficacy of the cognitive stimulation cannot be guaranteed. A simple example of this is a failure to cognitively stimulate, because the subject did not hear the instructions. A more realistic example may be the failure of DLPFC activation during a verbal fluency task may reflect an abnormal perceptual set or attentional processing in some schizophrenic patients. This failure would result in attenuated activation and poor performance, but not due to regional dysfunction or an intentional deficit. It is therefore likely that the more noteworthy cognitive activation studies using PET in schizophrenia will include more than one task. These tasks will, at some level, reveal behaviourial and neurophysiological responses that are dissociable in terms of a failure to form motor [intentional] set and perceptual set.

As noted above there are stimulations, such as pharmacological challenge, which are less complicated when it comes to interpretation. However even here it is likely that cognitive stimulation will have a role.

Neuropharmacology

It is probable that the key pathophysiology in schizophrenia is neurotransmitter related. Although integration of information in the CNS is determined by anatomically addressed connectivity, the mediation and modulation of information processing depends on neurotransmitters (Mesulam, 1990; Servan-Schreiber et al., 1990). The effect of pharmacological manipulation is likely to be a cornerstone for the understanding of pathophysiology in schizophrenia. The functional impact of altering neurotransmitter function can be measured using PET in conjunction with pharmacological activation.

For example buspirone, a partial agonist at 5-HT_{1A} receptors brings about a hypofrontality (Friston et al., 1991d) reminiscent of the rCBF profiles associated with psychomotor poverty and the activation profile brought about by verbal fluency (see Fig. 1c).

Study 3 — A psychopharmacological study

We have examined effects of buspirone (a 5-HT_{1A} receptor partial agonist) on changes in regional cerebral blood flow (rCBF) associated with free word recall.

Six subjects were scanned six times, using a 2×3 layout, with three pairs of memory tasks. A dose of 30 mg of buspirone was given orally after the first pair. Each memory task pair comprised a baseline subspan memory task and a supraspan task. The subspan task involved nine presentations of a five-word list with immediate free recall. The supraspan consisted of three presentations of a 15-word list, again with free recall. Presentation rate was one word per two seconds. The words chosen were high frequency, concrete words.

The critical difference between the subspan and supraspan tasks was the degree to which words have to be remembered. Only in the supraspan task is there a requirement to recall words that are not immediately accessible from short-term memory. Other components such as speaking, listening and attending were the same for both tasks.

The inter-pair spacing was 20 minutes. There was an eight minute interval between the subspan and supraspan task within each trial pair. The factorial design used allowed us to demonstrate a significant interaction (changes in rCBF brought about by psychological activation which were modulated by the drug) in the left parahippocampal region. This interaction was an attenuation of increases in local neuronal activity (rCBF) related to memory function. Buspirone-induced decreases in rCBF, independent of the memory effect, were seen in the left prefrontal and parietal cortices.

The direct effect of buspirone on DLPFC may not be attributable to 5-HT_{1A} receptor-buspirone interaction. Buspirone has an active metabolite 1-(2-pyrimidinyl)-piperazine (1-PP) which binds with nanomolar affinity to the alpha$_2$ adrenoreceptor where it acts as an antagonist (Bianchi and Garratini, 1988). There is evidence to suggest that alpha$_2$-receptors in the prefrontal cortex have a critical role in delayed response tasks. Clonidine (an alpha$_2$ agonist) reverses age-related deficits in delayed-response task performance in non-human primates (Arnsten and Goldman-Rakic, 1985) and this effect can be antagonized in a dose-dependent manner by the adrenergic antagonist yohimbine. Furthermore, pharmacological profiles in animals with lesions restricted to the DLPFC indicate that this area may be the site of action for some of clonidine's beneficial effects.

Buspirone also binds appreciably to dopamine receptors so an effect mediated through dopamine neurotransmission would not be excluded by these data.

Neuromodulation

Most of the neurotransmitters implicated in schizophrenia are neuromodulatory. Neuromodulation in this context means that the transmitter does not have a direct effect on transmembrane potential but will alter the responsiveness to independent input. This is important because the effects of manipulating neuromodulatory transmitter systems will only be evidenced by an attenuation or augmentation of an independent stimulation. A non-pharmacological (or non-modulatory pharmacological) challenge may therefore be necessary to "reveal" a neuromodulatory action of drug challenge. Indeed, the above factorial design was predicated on this argument. Short of magneto-stimulation, neuropsychological activations are the only way of delivering a non-pharmacological challenge to the brain regions implicated in schizophrenia. The choice of such independent challenges to neuronal activity will be crucial to interpretation. The difference between a schizophrenia cohort and a normal or other pathological group may, in this class of experiment, include formation of perceptual set, formation of intentional set, and neuromodulatory consequences of the drug challenge. Clearly, to demonstrate a specific difference in neuromodulatory response the formation of set must be equivalent for the two groups. The type of cognitive task chosen will be very different from those used in purely psychological studies.

Conclusion

The combined use of cross-sectional studies of schizophrenic groups and related activation studies in normal subjects is a powerful way to proceed. The multidimensional structure of schizophrenic symptom groupings, coupled with the distributed nature of cognitive brain systems complicates the observed relationships between regional neurophysiology and behavioural correlates. The three-way link between regional brain dysfunction, functional anatomy and behavioural deficits can only be established using activation studies in normal subjects. The extension of activation studies to patients is obvious but interpretation may be more complicated. The conjoint manipulation of two stimulations using factorial designs with PET has been introduced. An exciting area for future research may be combined psycho-pharmacological activations.

Acknowledgement

K.J.F. is supported by the Wellcome Trust.

References

Allen HA, Frith CD (1983) Selective retrieval and free emission of category exemplars in schizophrenia. Br J Psychol 74: 481–490

American Psychiatric Association (1987) Diagnostic and Statistical Manual of Mental Disorders, 3rd edn. American Psychiatric Press, Washington DC

Andreasen NC (1986) Comprehensive assessment of symptoms and history. College of Medicine, University College of Iowa, Iowa

Arndt S, Alliger RJ, Andreasen NC (1991) The distinction of positive and negative symptoms: the failure of a two-dimensional model. Br J Psychiatry 158: 317–322

Arnsten AFT, Goldman-Rakic PS (1985) Alpha$_2$-adrenergic mechanisms in prefrontal cortex associated with cognitive decline in aged nonhuman primates. Science 230: 1273–1276

Bianchi G, Garattini S (1988) Blockade of alpha$_2$-adrenoreceptors by 1-(2-pyrimidinyl)-piperazine (PmP) in vivo and its relation to the activity of buspirone. Eur J Pharmacol 147: 343–350

Bilder RM, Mukherjee S, Reider RO, Pandurangi AAK (1985) Symptomatic and neuropsychological components of defect states. Schizophr Bull 11: 409–419

Cowen PJ, Gadhui H, Godsen B, Kolakowska T (1985) Responses of prolactin and growth hormone to L-tryptophan infusion: effects in normal subjects and schizophrenic patients receiving neuroleptics. Psychopharmacology 86: 164–169

DeLisi LE, Holcomb H, Cohen RM, et al (1985a) Positron emission tomography in patients with and without neuroleptic medication. J Cereb Blood Flow Metab 5: 201–206

DeLisi LE, Buchsbaum MS, Holcomb H, et al (1985b) Clinical correlates of decreased anteroposterior metabolic gradients in positron emission tomography (PET) of schizophrenic patients. Am J Psychiatry 142: 78–81

Ferrier D (1875) Experiments on the brain of monkeys. Proc Roy Soc (Lond) 23: 409–430

Fox PT, Perlmutter JS, Raichle ME (1985) A stereotactic method of anatomical localization for positron emission tomography. J Comput Assist Tomogr 9: 141–153

Frackowiak RSJ, Lenzi GL, Jones T, Heather JD (1980) Quantitative measurement of regional cerebral blood flow and oxygen metabolism in man using ^{15}O and positron emission tomography: theory, procedure and normal values. J Comput Assist Tomogr 4: 727–736

Friston KJ, Passingham RE, Nutt JG, Heather JD, Sawle GV, Frackowiak RSJ (1989) Localization in PET images: direct fitting of the intercommissural (AC-PC) line. J Cereb Blood Flow Metab 9: 690–695

Friston KJ, Frith CD, Liddle PF, Lammertsma AA, Dolan RD, Frackowiak RSJ (1990) The relationship between local and global changes in PET scans. J Cereb Blood Flow Metab 10: 458–466

Friston KJ, Frith CD, Liddle PF, Frackowiak RSJ (1991a) Plastic transformation of PET images. J Comput Assist Tomogr 15: 634–639

Friston KJ, Frith CD, Liddle PF, Frackowiak RSJ (1991b) Comparing functional (PET) images: the assessment of significant change. J Cereb Blood Flow Metab 11: 690–699

Friston KJ, Frith CD, Liddle PF, Frackowiak RSJ (1991c) Investigating a network model of word generation with positron emission tomography. Proc Roy Soc (Lond) B244: 101–106

Friston KJ, Grasby PJ, Bench CD, Frith CD, Dolan RD, Cowen PJ, Liddle PF, Frackowiak RSJ (1991d) The neurotransmitter basis of cognition: psychopharmacological activation studies using PET. In: Exploring functional anatomy with positron emission tomography. Wiley, Chichester (CIBA Foundation Symposium 163) (in press)

Frith CD, Friston KJ, Liddle PF, Frackowiak RSJ (1991) Willed action and the prefrontal cortex in man. Proc Roy Soc (Lond) B244: 241–246

Goldman-Rakic PS (1986) Circuitry of primate prefrontal cortex and regulation of behaviour by representational memory. In: Mountcastle VB, Bloom FE, Geiger SR (eds) Handbook of physiology, V. American Physiological Society, Philadelphia, pp 373–417

Goldman-Rakic PS (1988) Topography of cognition: parallel distributed networks in primate association cortex. Ann Rev Neurosci 11: 137–156

Goltz F (1881) In: MacCormac WJW (ed) Transactions of the 7th International Medical Congress, vol 1. Kolkmann, London, pp 218–228

Gur RE, Resnick AM, Alavi A, et al (1987a) Regional brain dysfunction in schizophrenia. I. A positron emission study. Arch Gen Psychiatry 44: 119–125

Gur RE, Resnick SM, Gur RC, et al (1987b) Regional brain function in schizophrenia. II. Repeated evaluation with positron emission tomography. Arch Gen Psychiatry 44: 126–129

Ingvar DH, Franzén G (1974) Abnormalities of cerebral blood flow distribution in patients with chronic schizophrenia. Acta Psychiatr Scand 50: 425–436

Irle E (1990) An analysis of the correlation of lesion size, localization and behavioural effects in 283 published studies of cortical and subcortical lesions in old world monkeys. Brain Res Rev 15: 181–213

Lammertsma AA, Cunningham VJ, Deiber MP, Heather JD, Bloomfield PM, Nutt JG, Frackowiak RSJ, Jones T (1990) Combination of dynamic and integral methods for generating reproducible functional CBF images. J Cereb Blood Flow Metab 10: 675–686

Lueck CJ, Zeki S, Friston KJ, Deiber MP, Cope P, Cunningham VJ, Lammertsma AA, Kennard C, Frackowiak RSJ (1989) The colour centre in the cerebral cortex of man. Nature 340: 386–389

Liddle PF (1987) The symptoms of chronic schizophrenia: a re-examination of the positive-negative dichotomy. Br J Psychiatry 151: 145–151

Liddle PF, Friston KJ, Frith CD, Hirsch SR, Frackowiak RSJ (1991) Cerebral blood flow abnormalities associated with schizophrenic syndromes. Biol Psychiatry 29: 716S–717S

Mesulam MM (1990) Large scale neurocognitive networks and distributed processing for attention language and memory. Ann Neurol 28: 597–613

Mortimer AM, Lund CE, McKenna PJ (1990) The positive-negative dichotomy in schizophrenia. Br J Psychiatry 157: 41–49

Passingham RE, Chen YC, Thaler D (1989) Supplementary motor cortex and self initiated movement. In: Ito M (ed) Neural programming. Japan Scientific Societies Press, Tokyo, pp 13–24

Petersen SE, Fox PT, Posner MI, Mintun M, Raichle ME (1988) Positron emission tomographic studies of the cortical anatomy of single word processing. Nature 331: 585–589

Phillips CG, Zeki S, Barlow HB (1984) Localization of function in the cerebral cortex. Past, present and future. Brain 107: 327–361

Posner MI, Sandson J, Dhawan M, Shulman G (1989) Is word recognition automatic? A cognitive anatomical approach. J Cogn Neurosci 1: 50–60

Servan-Schreiber D, Printz H, Cohen JD (1990) A network model of catecholamine effects: gain, signal to noise ratio and behaviour. Science 249: 892–895

Spinks TJ, Jones T, Gilardi MC, Heather JD (1988) Physical performance of the latest generation of commercial positron scanner. IEEE Trans Nucl Sci 35: 721–725

Szechtman H, Nahmias C, Garnett S, et al (1988). Effect of neuroleptics on altered cerebral glucose metabolism in schizophrenia. Arch Gen Psychiatry 45: 523–532

Talaraich J, Tournoux P (1988) A co-planar stereotaxic atlas of a human brain. Thieme, Stuttgart

Weinberger DR, Berman KF, Zec RF (1986) Physiologic dysfunction of the dorsolateral prefrontal cortex in schizophrenia. 1. Regional cerebral blood flow evidence. Arch Gen Psychiatry 43: 114–124

Wise S (1989) Frontal cortex activity and motor set. In: Ito M (ed) Neural programming. Japan Scientific Societies Press, Tokyo, pp 25–38

Zeki S, Watson J, Lueck C, Friston KJ, Kennard C, Frackowiak RSJ (1991) A direct demonstration of functional specialization in human visual cortex. J Neurosci 11: 641–649

Zihl J, von Cramon D, Mai N (1983) Selective disturbance of movement vision after bilateral brain damage. Brain 106: 313–340

Authors' address: Dr. K. J. Friston, MRC Cyclotron Unit, Hammersmith Hospital, Ducane Road, London, W12 8HS, United Kingdom

Subject Index

A.H. Tuma, E. Stricker, S. Gershon (eds.)

Advances in Neuroscience and Schizophrenia

(Journal of Neural Transmission, Supplementum 36)

Schizophrenia has been the subject of intense research interest in recent years, as investigators have explored the biological bases for the disorder and for various approaches to its diagnosis and treatment. This volume focuses on three aspects of such recent research connecting basic neuroscience to schizophrenia. In one, Professors Budinger, Gur, and Pettegrew provide critical reviews of brain imaging studies as they relate to cognitive behavior functions in schizophrenia. In the second, Professors Goldman-Rakic, Lewis, and Tassin discuss monoamine systems and their varied role in prefrontal cortical circuitry and function. In the third, Professors Deutch, Gerfen, and Grace discuss the structure , organization, and function of the basal ganglia, as they relate to schizophrenia and the mechanisms of neuroleptic action. These papers were presented at an interdisciplinary workshop on the subject at the University of Pittsburgh, in May 1991, and the discussion between the authors and other panelists in basic and clinical sciences are included in this volume as well. The presentation of these diverse approaches in an integrated fashion provides the reader with a unique perspective and a wealth of new questions for future collaborative research.

1992. Approx. 40 figs. Approx. 160 pages.
Soft cover DM 110,-, öS 770,-
Reduced price for subscribers to
"Journal of Neural Transmission":
Soft cover DM 99,-, öS 693,-
ISBN 3-211-82347-6

Springer-Verlag
Wien New York

Prices are subject to change without notice

Chr. Marescaux, M. Vergnes, R. Bernasconi (eds.)

Generalized Non Convulsive Epilepsy: Focus on GABA-B Receptors

(Journal of Neural Transmission, Supplementum 35)

Generalized non convulsive epilepsy (GNCE) also called absence or petit mal epilepsy, is a disease appearing during childhood in humans. EEG, clinical, pharmacological and genetic characteristics differ from those of convulsive or focal epilepsies. No underlying structural or biochemical abnormality has been identified for generalized absence seizures and the etiology of this disorder is unknown. It is unlikely that the precise pathophysiology of GNCE can be resolved by the study of human subjects. Therefore a number of animal models reproducing the human disease have been developed. The aim of this supplementum is to characterize such models in rodents. First, recent models are extensively described. These include the genetic model of spontaneous GNCE in Strasbourg´s Wistar rats and in tottering mice as well as bilateral spike and wave discharges induced by GHB, PTZ or GABA mimetics.

Second, this supplementum will also provide very recent information on putative mechanisms underlying generalized absence seizures. Third, various experimental approaches aimed at investigating the neural substrate of this particular kind of epilepsy are described with reports of various electrophysiological, pharmacological, biochemical, metabolic, ionic and molecular data. The supplementum provides an original multidisciplinary approach to the mechanisms involved in GNCE and demonstrates that rodent models constitute a promising tool which complements the classical feline penicillin model.

Springer-Verlag
Wien New York

1992. Approx. 61 figures. Approx. 160 pages.
Soft cover DM 110,-, öS 770,-
Reduced price for subscribers to
"Journal of Neural Transmission":
Soft cover DM 99,-, öS 693,-
ISBN 3-211-82340-9

Prices are subject to change without notice